CAREERS
IN HORTICULTURE
AND BOTANY

JERRY GARNER

VGM Career Horizons
NTC/Contemporary Publishing Group

Cover photo of Chin-chang Chu pollinating Phalenopsis orchid courtesy of Dr. Kenneth W. Mudge, Department of Floriculture and Ornamental Horticulture, Cornell University. Copyright KWMudge, 1995.

Library of Congress Cataloging-in-Publication Data

Garner, Jerry L.
  Careers in horticulture and botany / by Jerry L. Garner.
    p.   cm.
  Includes bibliographical references (p.   ).
  ISBN 0-8442-4458-9 (hardcover : alk. paper). — ISBN 0-8442-4459-7
(pbk. : alk. paper)
  1. Botany—Vocational guidance.   2. Horticulture-Vocational
guidance.   I. Title.
QK50.5.G37   1996
580'.23—DC20   96-26669
                                                                                CIP

Published by VGM Career Horizons
A division of NTC/Contemporary Publishing Group, Inc.
4255 West Touhy Avenue, Lincolnwood (Chicago), Illinois 60646-1975 U.S.A.
Copyright © 1997 by NTC/Contemporary Publishing Group, Inc.
Printed in the United States of America
International Standard Book Number: 0-8442-4458-9 (cloth)
                                                    0-8442-4459-7 (paper)
18   17   16   15   14   13   12   11   10   9   8   7   6   5   4   3   2

CONTENTS

ABOUT THE AUTHOR

After ten years as a professor of horticulture and chairman of a horticulture department in the Virginia Community College System, Dr. Jerry Garner joined the staff of Mariani Landscape in Chicago, Illinois, as their Horticulturalist and Plant Buyer. In this position, he brought his extensive academic and consulting expertise in residential and commercial plant materials and garden design to one of the Midwest's most prestigious design/build firms specializing in estate work.

In addition to a bachelor's degree in English and graduate work in education from The College of William and Mary, Dr. Garner holds a bachelor's degree in horticulture with a specialization in floriculture from Virginia Polytechnic Institute and State University (Virginia Tech). Prior to his bachelor's work in horticulture, Dr. Garner was employed as a horticultural therapist and was extensively involved in the professional association in this area. After receiving his bachelor's degree, he pursued a Ph.D. in horticulture at Virginia Tech.

In Virginia, Dr. Garner served as the horticultural advisor to major businesses, government agencies, public universities, and industry executives. He also served on numerous horticultural advisory boards including the Virginia State Fairground Horticulture Pavilion and the Maymont Foundation and Park. In addition, he was a regular lecturer at the Lewis Ginter Botanical Gardens.

Dr. Garner was a founding board member of the Maymont Flower and Garden Show in Richmond, Virginia, and served on the executive board of the Virginia Society of Landscape Designers. Since coming to Illinois, Dr. Garner has become an active member of the Illinois Landscape Contractors Association, serving on the Education Seminars Committee. He is also on the Board of Directors of the Evanston Arts Center where he serves on the Exhibition, Benefits, and Nominating Committees.

ACKNOWLEDGMENTS

I would like to thank the many people I have worked with in the fields of botany and horticulture throughout my career for their inspiration for this book. The initial concept for this book grew out of their enthusiasm for their work and their genuine desire to encourage others to explore these fields as career options.

I would like to acknowledge the encouragement I received from my daughter, Lauren, and her numerous contributions to this book. Her academic and professional experience in the fields of botany and horticulture provided unique insight into both fields.

Special thanks go to my wife Gerry for first encouraging me to write this book, and for all of her love, patience, encouragement, and support in bringing it to completion.

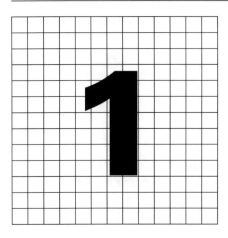

INTRODUCTION

ABOUT THIS BOOK

Plants are central to the ecological system of our planet as well as the aesthetic well-being of humans. Consequently, the study and the use of plant material has been important from the beginning of time. As food crops, plants have sustained life. As building materials, plants have protected people from the elements. As medicine, plants have soothed and cured many ailments that have afflicted humans.

To most people the term *plants* can mean many different kinds of living organisms ranging in size and complexity from microscopic bacteria to giant trees such as redwoods. Most laypeople include algae, fungi, mosses, ferns, cone-bearing plants, and flowering plants in the definition. Today most scientists place bacteria, algae, and fungi in their own distinct kingdoms, but many university and college plant science departments still include these groups in their courses.

In this century, basic research and information about plant life has increased so significantly that the results of this knowledge have been applied to more areas of our life. In addition, new career areas have evolved that offer opportunities for those interested in plants to pursue careers in a wider range of areas. For example, new advances in plant genetics have led to new fields of biotechnology in horticulture and forestry. Advances in plant pathology, the study of plant-related diseases, have resulted in new fields of disease management and control in greenhouses, nurseries, and forests. It is no wonder that the study of plants, from the cellular level to the entire plant, provides such a wide variety of career options.

One

Because of the many career options that are now available in botany and horticulture, this book is designed to

- Provide an overview of the career options in botany and horticulture;
- Define the occupational areas in botany and horticulture;
- Answer questions about pursuing various career options in the fields of botany and horticulture.

Some of the questions this book is designed to answer are

What are the differences between botany and horticulture?

What are the occupational options that I can pursue in either botany or horticulture?

How do I know which area(s) of botany and horticulture is right for me?

Where are people in botany and horticulture employed?

What are the earning capabilities of people in botany and horticulture?

How should I prepare for a career in botany or horticulture?

Where can I study botany and horticulture?

This book will concentrate on the career opportunities in the fields of botany and horticulture, as well as in some fields that are not easily categorized as botany or horticulture careers because their knowledge base is primarily botany but their application is primarily horticulture.

AN OVERVIEW OF BOTANY AND HORTICULTURE

Botany

The term *botany* is derived from the Greek *botane,* meaning plant or herb. Botany and its many branches are pure sciences dealing with the fundamental principles of plant life. In general, botanists study the overall development and life processes of plants. These include plant physiology, heredity, distribution, anatomy, and morphology, the environment, and the economic value of plants for application in other fields. The results of the basic research done by botanists are applied to four major fields of study: horticulture, agronomy, forestry, and pharmacology.

Horticulture is the study of the cultivation of plants. *Agronomy* is the study of the practical use of plant and soil sciences to increase the yield of crops. *Forestry* is the study of forest management for conservation and production of timber. And *Pharmacology* is the study of medicinal actions of substances.[1]

Botanists study the chemical composition of plants and they understand complex chemical combinations and reactions involved in metabolism,

---

1. Agronomy, forestry, and pharmacology each offer a vast array of career opportunities, which can be read about in such publications as VGM's *Careers for Plant Lovers and Other Green Thumb Types, Careers Encyclopedia, Dictionary of Occupational Titles,* and *Occupational Outlook Handbook.*

reproduction, growth, and heredity. They also study the behavior of plant chromosomes and reproduction, internal and external structures, and the mechanics and biochemistry of plants and plant cells. Also they may investigate plant environments and plant communities and the effects of rainfall, temperature, climate, and soil on plant growth, development, and evolution. Some botanists identify and classify plants. Some conduct environmental studies and prepare reports. Most often botanists are designated according to their field of specialization.

The results of botanical study can improve the quality of our medicines, food products, fibers, and building materials. These results may also help to improve our management of parks, wilderness areas, forests, or wetlands. An understanding of plant science may even lead to better solutions to problems caused by overpopulation and pollution.

Botanists are not necessarily concerned with applying their ideas outside the realm of research in the laboratory or the field. They may be most concerned with understanding the basic principles and laws governing plant structure, biochemistry, and function. Other branches of plant science, such as horticulture, might then apply these fundamentals to the production and utilization of plants.

Horticulture

By definition, *horticulture* is "the intensive cultivation of plants." The name is derived from the Latin terms *hortus* and *cultura,* meaning "garden cultivation." Today the term *horticulture* has evolved to refer to the art and science of growing and utilizing fruits, vegetables, flowers, and ornamental plants for the benefit of people and the environment. The intensity of crop cultivation distinguishes horticulture from most other agronomic sciences that are involved in the production of crops such as tobacco, rice, field corn, timber, and others. Horticultural crops such as roses must be regularly watered, pruned, fertilized, and sprayed for damaging pests. Loss of a few plants during production could result in economic loss to the grower. However, a corn crop, for example, is dependent upon the weather for watering, and it requires minimum fertilization prior to planting and minimum spraying for pests—and only if a major reduction in yield is imminent. In this case, the loss of a few corn plants would not result in any real economic loss to the grower.

Another distinction between horticultural crops and other agronomic crops is the intended use of the crop. For example, pine trees grown as a horticultural crop are carefully pruned, fertilized, and monitored to produce an aesthetically pleasing tree, each tree bringing a high price in the nursery and landscape market. However, pine trees grown for timber need not be aesthetically pleasing, thus their cultivation requirements are less demanding, and each individual tree would not command as high a price for the grower.

Typically, horticulturists are concerned with the study of so-called higher plants rather than lower plants. Higher plants are plants that reproduce from seed produced by flowers or cones. Examples would include petunias and pines. Lower plants reproduce by spores or by breaking apart into pieces that then grow into another plant. Examples include ferns or mosses. There are exceptions to this distinction, however. For example, ferns are typically grown in greenhouses and in perennial plant nurseries, both of which are the workplace of the horticulturist. Horticulturists most often work with entire plants rather than plant cells, and they are more concerned with plant production than with the fundamental cellular composition, biochemistry, and classification of plants.

Horticulture is an applied science, relying upon many other disciplines such as chemistry, physics, engineering, art, meteorology, economics, entomology, botany, and many more. A horticulturist uses the basic information and theories of these related fields and tries to find an application for these ideas for the benefit of people and the environment.

Botany and horticulture are interrelated but distinct branches of science. Both involve the scientific study of plants, but on different levels and for different purposes. This is not to imply that one career field is more difficult or more important than the other. Both provide a range of opportunities requiring various levels of education and offering various levels of compensation. Because both fields are so broad, each has a wide variety of career choices with an equally wide variety of career paths a person could follow to reach his or her career goals.

As you can see, the terms *botanist* and *horticulturist* are very general career titles. Rarely will you find a person in either of these fields who has not specialized somewhat and narrowed down his or her work into one of the subdivisions of these two careers. These terms are simply umbrella titles under which many diverse and more specialized careers can be found. Following are general overviews of botany and horticulture careers. More detailed information can be found in the section on each specific career.

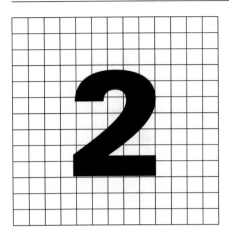

CAREERS IN BOTANY

Plants are not only complex living organisms but also chemical factories. They produce many chemicals that are useful to humans. In fact, according to Weier, Stocking, Barbour, and Rost, "plants are the only producers. All consumers, particularly people, are dependent upon plants for food, fiber, wood, energy, and oxygen." In addition to producing food, plants also produce the raw materials for paper, building materials, solvents and adhesives, fabrics, medicines— some used to treat certain types of cancer—and many other products.

In general, botanists study the chemical composition of plants and they understand the complex chemical combinations and reactions involved in the metabolism, reproduction, growth, and heredity of plants. The nature of botanists' work depends to a large degree on the area in which they specialize and the setting in which they conduct their work.

SPECIALIZATIONS IN BOTANY

According to the Botanical Society of America (BSA) the specializations within the field of botany can be divided into three major categories with numerous disciplines related to each specialization. The BSA specializations and their related disciplines of study include plant biology specialties, applied plant specialties, and organismic specialties.

Plant Biology Specialties Plant biology specialties include the following:

Anatomy—microscopic plant structure (cells and tissues).

Biochemistry—chemical aspects of plant life processes, including the chemical products of plants.

Biophysics—application of physics to plant life processes.

Cytology—structure, function, and life history of plant cells.

Ecology—relationships between plants and the world in which they live.

Genetics—plant heredity and variation. Plant geneticists study genes and gene function in plants.

Molecular biology—structure and function of biological macromolecules. Molecular biology includes biochemical and molecular aspects of genetics.

Morphology—macroscopic plant form related to the evolution and development of leaves, roots, and stems.

Paleobotany—biology and evolution of fossil plants.

Physiology—functions and vital processes of plants. Photosynthesis and mineral nutrition are two examples of subjects studied by plant physiologists.

Systematics—evolutionary history and relationships among plants.

Applied Plant Specialties

Applied plant specialties include the following:

Agronomy—crop and soil sciences where practical applications of plant and soil sciences are used to increase the yield of field crops.

Biotechnology—using biological organisms to produce useful products. Plant biotechnology involves inserting desirable genes into plants and having those genes expressed.

Breeding—development of better types of plants.

Economic botany—the study of plants with commercial importance, including harmful and beneficial plants and plant products.

Food science and technology—development of food from various plant products.

Forestry—forest management for the production of timber and conservation.

Horticulture—the production of ornamental plants and fruit and vegetable crops. Landscape design is an important subdiscipline in horticulture.

Natural resource management—the responsible use and protection of natural resources.

Plant pathology—diseases of plants, including the biological aspects of disease and disease management or control.

Organismic Specialties

Bryology—the study of mosses and similar plants.

Lichenology—the biology of lichens, dual organisms composed of both a fungus and an alga.

Microbiology—the study of microorganisms.

Mycology—the study of fungi.

Phycology—the study of algae.

Pteridology—the study of ferns and similar plants.

However, Weier, Stocking, Barbour, and Rost, in their book, *Botany,* provide a more concise method of identifying the five basic problems that concern botanists. They are:

1. Naming and classifying plants;
2. Explaining the principles underlying the diversity of plant life;
3. Identifying the mechanisms that control the precise patterns of growth;
4. Identifying the types of plants that grow under different conditions of temperature, soil, moisture, and nutrients in order to determine which plants will produce most abundantly; and
5. Studying the relationship of plants to their environment in order to determine how harmful changes in the environment can be reversed.

Naturally, these issues relate to the five major career fields in botany: plant taxonomists, plant ecologists; plant morphologists, plant physiologists, and plant cytologists. In the next section of this book, these five occupational areas in the field of botany will be described in terms of the work performed and then in terms of the work environment, employment outlook, preparing for a career in botany, other desirable personal qualifications, and salary and other benefits.

THE WORK PERFORMED BY BOTANISTS

The work of botanists varies depending on which of the five major areas of botany they have specialized in. To gain a better understanding of the work of a botanist, it is best to know the nature of the work that can be performed in each specialization.

In general botanists study the development and life processes of plants. They also study the physiology of plants (the functions and life processes of plants such as photosynthesis and nutrition); the anatomy of plants (plant cells and tissues); the morphology of plants (the evolution and development of leaves, roots, and stems); plant heredity; the environment; plant distribution; and the economic value of plants.

Overall, botanists study the behavior of plants from the chromosome level to the reproduction process. They also study plant structures that are internal (cellular) and external (leaves and stems). In addition, botanists study the mechanics of plants (the structure of plants and how they work) and the biochemistry of plants (the chemical aspects of plant life, including the chemical products of plants and/or plant cells).

Following is a brief overview of specific careers within the five major areas of botany.

PLANT TAXONOMIST

Plant taxonomists are botanists who work to identify, describe, classify, and name plant species. In so doing they provide all plant scientists with a consistent means of communication. The importance of this work is readily understood when one considers the importance of knowing the correct name, description, and classification of a plant species in almost any field within botany and horticulture. For example, in the research work of the plant physiologist knowing the exact classification of a plant species is essential to understanding the physical characteristics of a specific plant. For the nursery grower, this knowledge is central to using the correct propagation and growing techniques as well as knowing how to sell the plant material for proper installation in a landscape design. Likewise, the floral designer must know the classification and characteristics in order to use a particular flower or plant in an arrangement designed for certain conditions.

In addition to attempting to inventory the earth's plant resources, plant taxonomists also identify new plant species previously unknown to the field. Such discoveries and identifications may potentially lead to the identification of new food crops or new drugs which can cure or treat human diseases.

In addition, the work of plant taxonomists may also produce a new source of plant genes. The identification of new gene sources may ultimately improve the survival of endangered plant life. By using the discoveries of plant taxonomists, plant geneticists may be able to make endangered plant species stronger and more resistant to predators.

The Work Environment

Plant taxonomists are generally not able to work regular hours. They must coordinate their work with the seasonal demands of the plant materials they are studying. In some cases, they may be exposed to unsafe or unhealthy working conditions but they take safety precautions when working under such conditions.

Most plant taxonomists are engaged in field work. This type of work involves strenuous physical activity and sometimes less than comfortable or safe living conditions. This is particularly true if their research and work is conducted in areas such as rain forests, deserts, or countries with primitive or unstable living conditions. Plant taxonomists must enjoy the outdoors and commonly travel to remote places.

Some of the tasks performed by plant taxonomists involve working in a laboratory and operating very sophisticated equipment and following strict scientific protocol procedures. Plant taxonomists are employed by colleges and universities as well as by such organizations as botanical gardens and arboreta.

PLANT ECOLOGIST

Plant ecologists are botanists who work to understand the relationships between plants and the world in which they live. Because these interactions are complex, the plant ecologist studies the influences of such things as population size, pollutants, rainfall, temperature, and altitude on plant organisms. In some respects, the work of plant ecologists is like that of plant taxonomists. Plant ecologists inventory plant resources. However, unlike the taxonomists, plant ecologists categorize groupings of plant species. They are interested in learning the history of these plant communities and how their paths across the world changed over time.

Plant ecologists examine such things as the environmental factors that regulate plant growth. They also study the responses of plants to favorable and unfavorable environmental conditions. In many cases, plant ecologists attempt to understand the natural balance between plants and their environment. In other cases, they attempt to manipulate this balance to either better understand the relationship or to achieve a better result.

One of the overriding concerns of plant ecologists is determining the most efficient methods of using plant resources. In order to do this, plant ecologists draw on their comprehensive knowledge of plant taxonomy, physiology, and anatomy as well as their knowledge of geology and meteorology. Plant ecologists must also rely on their strong background in chemistry.

The Work Environment   Some plant ecologists may be able to work more regular hours than plant taxonomists; however, they too must coordinate their work with the seasonal demands of the plant materials and with the organizational demands of their employer or client. This is particularly true when environmental impact studies are needed in advance of or in response to construction projects.

In some cases, plant ecologists may be exposed to unsafe or unhealthy working conditions. When plant materials have to be studied after an environmental disaster, for example, the conditions in which the plant life is found may not be healthy or safe for the botanist. However, they take safety precautions when working under such conditions.

Most plant ecologists are engaged in field work. The work can involve strenuous physical activity as well as more sedentary laboratory work. Some of the tasks performed by plant ecologists involve data collection, statistical analysis, and the use of sophisticated monitoring or measurement equipment. Plant ecologists are employed by government regulatory agencies and departments as well as by colleges and universities. Some may also be employed by nonprofit environmental organizations.

PLANT MORPHOLOGIST

From what you have read so far, and what you have observed in nature, you know that plant life comes in all forms, from the most microscopic organisms to the splendor of majestic trees. Botanists who have an interest in why and how these plant species differ are called plant morphologists.

Plant morphologists study the manner in which cells are arranged to give each plant its form. Their interests lie in the wide range of plant forms that exist in nature. To determine the cellular arrangement of plants, plant morphologists study how different plants transport and conduct food and water.

Considering the differences between cacti and seaweed, for example, one can readily see how differently the cells meet the requirement of supplying these plants with nutrients and water.

A branch of plant morphology known as *morphogenesis* is the study of the reproductive structures of plants to determine the mechanisms involved in how a particular variety develops. Plant morphologists know that cells divide in precise planes and enlarge in precise locations within a developing seed at regular rates. In some species, they have learned to modify some of these steps and would like to know more about the manner in which the steps are controlled and directed within the plant itself. The true value of the work of the plant morphologist is that his or her work will provide increased knowledge about all cells.

The Work Environment   In general, plant morphologists work in research laboratories. However, they are like almost all other botanists in that their work must involve field work and consequently their working hours may sometimes have to meet the demands of the plant species they are researching. For example, conducting research on the reproductive process of a particular plant form will require different conditions for pollination and germination. The very nature of this process may dictate when the researcher must be available to make observations or apply treatments.

Plant morphologists are less likely to be exposed to unsafe or unhealthy working conditions than plant taxonomists and plant ecologists. However, their research may require them to work with chemical products that necessitate strict safety procedures.

Plant morphologists are likely to be engaged in field work as well as laboratory work. Therefore, their work can be quite strenuous. Laboratory work often involves many hours spent analyzing slides under various types of microscopes or using special techniques for analyzing the cellular structure of the plant material. In addition, these botanists use a variety of monitoring and measurement devices to collect their data. The data collection then requires statistical analysis and report working.

Plant morphologists may be employed by government agencies and private industry, particularly the timber industry. They are employed in other organizations such as botanical gardens and arboreta as well. The majority of plant morphologists are employed in colleges and universities.

PLANT PHYSIOLOGIST

Plant physiologists are botanists who study the internal functions and processes of plants. The process most well known to the layperson is photosynthesis. In addition to photosynthesis, plant processes also include growth, respiration, circulation, excretion, movement, reproduction, and other functions of plants. In order to study these processes plant physiologists conduct research on the cellular structure as well as organ-system functions of plants. They have two major goals in their research.

The first goal of the plant physiologist is to develop a better understanding of the biological, chemical, and physical processes that are basic to plant life. The second goal is to regulate and control these activities in order to shape plant growth and development. Because the basic processes govern the diversity of plant life, by understanding these processes, it is possible to learn to control when certain plants will flower. For example, by controlling the environment to regulate certain processes, growers can assure that lily plants will be in bloom for Easter or that bedding plants will flower at the appropriate time. Likewise, fruits and vegetables can be controlled to give them certain characteristics that will increase their value in the market place.

Plant physiologists conduct their studies of plant processes under either normal conditions for the species or abnormal conditions. This decision is determined by the nature of the study and by the effects of the internal and external environmental factors on the life processes of plants.

The Work Environment   Some plant physiologists work regular hours in a research laboratory. However, most plant physiologists are like other botanists. They must coordinate their work with the demands of the plant materials and the research requirements. Consider the wide variety of growing conditions for various species of plants. In order to conduct research on the reproduction process certain periods of dormancy must be simulated regardless of the season or the time of day. This may require the researcher to cover and uncover plant specimens at various times during the day or night—and to do so seven days a week.

Like plant morphologists, plant physiologists are less likely to be exposed to unsafe or unhealthy working conditions than some other botanists. However, their research may require them to work with chemical growth regulators or other chemical products that require strict safety procedures to be followed.

Many plant physiologists are engaged in field work as well as laboratory work. Like other botanists, their work can be strenuous as well as sedentary. Laboratory work often involves time spent analyzing slides under various types of microscopes or using precise staining techniques. In addition, plant physiologists use a variety of monitoring and measurement devices to collect their data. The data collection then requires statistical analysis and report writing.

Plant physiologists may be employed by government agencies, private industry, and organizations such as botanical gardens and arboreta. However, the majority are employed on the faculties of colleges and universities.

PLANT CYTOLOGIST

Botanists who study plant cells are known as cytologists. They study the structure, function, and life history of plant cells. Cytologists select and section minute particles of plant tissue for microscopic study because they are concerned with phenomena at the cellular level. For example, if the genetic code present in the nucleus of a cell directs the synthesis of a particular protein, the cytologist wants to know the exact relationship between the steps in the photosynthesis and respiration and the molecules of the membranes associated with this process. They are concerned with the physical arrangement of DNA in a particular species of plants.

Some of the other areas cytologists study include the cells concerned with reproduction as well as the means by which chromosomes divide or unite. They also study the formation of sperm and eggs in a variety of plant species and the physiology of unicellular organisms. These studies ascertain physical and chemical factors involved in growth of the plant material.

Cytologists also evaluate exfoliated, aspirated, or abraded cells or assess the hormone status and the presence of atypical or malignant changes in various plants. Because cytologists are concerned with plant tissue culture, they are able to grow entire plants from single cells in some laboratories.

The cytologist's research has the potential to expand biotechnology fields in horticulture, forestry, and plant pathology. For example, botanists working in this area might develop a blight-resistant strain of a species of tree from a single cell. This means that endangered trees which have been afflicted by disease or fungus can be propagated very quickly.

Because of the relative ease in studying plant cells, as opposed to animal cells, the research of cytologists has and will continue to contribute to the body of knowledge necessary to make advances in fields such as pharmacology.

The Work Environment

Cytologists tend to work in research laboratories. Although they do not tend to conduct field work to the same extent as plant taxonomists and plant ecologists, they coordinate some aspects of their work with the demands of the plant materials in the field.

Like plant morphologists and plant physiologists, cytologists are less likely to be exposed to unsafe or unhealthy working conditions. When their work does require them to be exposed to various chemicals, they follow strict safety procedures to assure their own safety and that of the technicians in their laboratories.

Laboratory work is often sedentary and individualistic. Because the study of cells involves considerable time analyzing slides with various types of instruments and under various types of conditions, the work has to be precise. Cytologists employ a variety of techniques, such as staining, to make cell structures visible or to differentiate parts. Monitoring and measurement devices have to be calibrated to assure that the data collected during these research studies is accurate. Statistical analysis and report writing are standard aspects of the work of cytologists.

Cytologists are employed by government agencies and private industry, particularly the pharmaceutical industry. However, the majority are employed at land grant and research universities.

EMPLOYMENT OUTLOOK FOR BOTANISTS

Demand for botanists is expected to increase faster than average through the turn of the century, particularly for those with advanced degrees, certain specialized degrees, and certain areas of experience. This trend should continue into the next century with the most growth in private industry. Only 10 percent of all botanists are employed in private industry. Drug companies, chemical companies, the oil industry, food companies, and lumber and paper companies are but a few of the major industrial employers of botanists. Areas of job growth will occur also in the genetic research industry, where the search for new drugs and medicines and useful genes for improving food crop plants will continue to create a demand for botanists interested in both laboratory study and in botanical exploration. Botanists who focus on environmental processes and problems will also continue to be in great demand by industry as public concerns for the environment increase in areas such as air, soil, and water pollution.

Approximately 40 percent of botanists work in government agencies. Federal and state agencies will continue to need botanists in many different areas. The U.S. Department of Agriculture will increase its demand for botanists with training and experience in areas such as medicinal plant research, plant diseases, and plant genetics and breeding. The U.S. Department of the Interior employs botanists with interests and training in plant ecology and conservation. Even agencies such as the State Department, the Smithsonian Institution, the Environmental Protection Agency, and the National Aeronautics and Space Administration will continue to have a need for qualified botanists. In addition to the federal agencies listed above, all 50 state governments employ botanists in positions similar to those in the federal government. Slower growth than in industry can be expected due to current political and governmental priorities and possible budget cuts, but job stability will probably be good because of employment on long-term research projects.

Academic institutions will continue to be the primary employers of botanists, ranging from high schools, technical training schools, and community colleges to four-year colleges and universities. More openings will exist in high schools, technical schools, and community colleges for general botanists—that is, those less interested in specialized areas of botany and in research. Most institutions at this level will need people who can teach a wide range of plant-related courses, and time and equipment for research in these schools are usually very limited. Over 50 percent of all botanists are employed in higher education institutions. Four-year colleges and universities will continue to need botanists with a wide variety of specializations. Almost all of these schools offer courses and degrees in botany or plant science; thus there are faculty positions for botanists who have different specialties. Also, research and administrative positions are common in these educational institutions. Just as in government positions, slower growth may be expected in academia than in industry due to political and social trends and possible budget cuts that affect the availability of funds for both teaching and research positions in education. Job stability, however, will remain good due to long-term projects and the continuing need for qualified faculty.

Many other careers for botanists do not involve teaching or research. Some botanists work in administration and marketing for biological supply houses, seed companies, biotechnology firms, pharmaceutical manufacturers, and scientific publishers. Some work as scientific writers, illustrators, and photographers.

Any economic recession would tend to limit opportunities for botanists. Employment opportunities will vary depending on local, state, and national economies. Some specializations may be in demand more than others, and any oversupply of persons with training in a particular field will make competition for positions greater. However, the field of botany is so diverse that challenging positions will be available for any well-trained botanist.

PREPARING FOR A CAREER IN BOTANY

High School Preparation   Students in high school should take a college preparatory curriculum including courses in English, mathematics, foreign language, physics, chemistry, and biology. Courses in social science and the humanities will also prove valuable for later college-level studies. Any courses that will help the student develop good communication skills should be taken also.

If you are planning to pursue a bachelor's degree in botany or horticulture, your high school curriculum should include the following courses:

- Algebra I and II
- Trigonometry
- Biology
- Physics
- Social studies (3 units)
- Fine arts or humanities (1 or 2 units)
- Computer programming or computer applications
- Geometry
- Calculus
- Chemistry
- English (4 units)
- Foreign languages (2 or 3 units)

Other courses that may be helpful include economics, history, and public speaking. It is also recommended that prospective botany and horticulture students take advanced placement or honors level courses. It is advised that you set a goal of achieving combined scores of at least 1000 on the SAT exam or 20 on the ACT exam.

Admissions officers at colleges and universities also look for well-rounded students. Extracurricular activities during high school can reflect this. Being a member and holding an office in science clubs will demonstrate strong and consistent interests. However, participation in athletics, service organizations, and cultural activities are also important.

Extracurricular activities that will prove valuable include participation in science clubs and science fairs. Summer jobs and internships in any area related to biology also help. This might include working in parks, laboratories, public and private gardens, plant nurseries and greenhouses, farms, and camps. Hobbies such as gardening, photography, and camping as well as computer skills are also useful.

Prior to applying to any college or university the student should get as much information as possible from each school being considered. Your counselor and library can also help. Many schools do not have separate botany departments but instead teach botany in a department of biology. In any case write, call, and visit, if possible, the schools that interest you, and ask to meet with some of the botany faculty to discuss your career goals and options and how they feel their department will help you fulfill your goals. The Botanical Society of America maintains a list of botanists throughout the country, with their addresses and phone numbers. Contact some of these botanists for advice. The address of the Botanical Society of America is listed in the appendix at the end of this book.

College Curriculum

Bachelor degree programs in biology and chemistry are readily available through almost all colleges or universities. Botany programs at the undergraduate level are less readily available. George Washington University in Washington, D.C., is one of the few universities to offer B.S. degrees in botany. Horticulture programs are available at every land grant college in the United States.

The specific courses you take in college will vary depending on the particular curriculum of the college you attend and on your own interests. Most colleges and universities require a core program in biology before you can enroll in any specialized courses in botany. At other colleges you may be able to take botany courses immediately.

In general, the first two years concentrate on mathematics and the physical sciences with introductory biology and chemistry courses as well as courses in English and the social sciences. The last two years include required courses in the major. In addition, students take required electives, which help focus the undergraduate degree in your major, and "free" electives. "Free" electives are courses that the student wishes to take. These courses can either be any course outside of your major department or any course on a list of electives approved by your department.

To be best prepared for the job market or for graduate school, you should take courses in writing and literature, the arts, humanities, and social sciences in addition to your general biology and specialized botany courses. Most curricula will also require that you take courses in general math through calculus, statistics, chemistry, and physics. You should also become proficient in using a computer and a wide variety of software programs. Some schools may require a speaking and writing ability in a foreign language. This could be especially important for a field botanist interested in working in a location other than the United States.

Summer jobs, internships, and cooperative education can provide important work experience for students. Positions exist in government, college and university laboratories, agricultural and biological research stations, and in private industry.

If at all possible, arrange to do an undergraduate research project under one of your professors. This might be an independent research project of your own or assisting the faculty member with his or her research. This type of experience can help you decide which areas of botany you like best or least, and it will give you invaluable experience and insight into how science works. Obviously it can also help you decide whether graduate school and a future as a research botanist is the best choice for you.

Other Personal Qualifications

Botanists must have clear and concise verbal and written communication skills. They should have a high level of curiosity about the world around them. They need to be creative in solving problems but at the same time have a clear understanding of scientific method and the rigors of scientific research. Generally, botanists should be detail oriented, careful and precise in their work. Some specialties in botany require physical strength and stamina; other specialties may be sedentary. All areas require mental stamina due to long hours in the laboratory and in interpreting data. Most botanists would benefit from good interpersonal skills for collaborative work and interaction with colleagues and students. Some specialties, however, may be better suited to the person who prefers to work alone with only limited interaction with others.

Botanists are often required to understand and operate a variety of scientific equipment ranging from the simple to the extremely sophisticated. In today's scientific world computer skills are a necessity. A clear understanding of a variety of software programs is a must and some skills in programming would be beneficial. The ability to work with microscopes ranges from the simple microscopes found in most high school biology laboratories to sophisticated electron microscopes, common today in most university and industry laboratories. Other beneficial skills might include staining techniques, radioisotope analysis, digital imaging analysis, satellite imaging, telemetry, and cell and tissue culture. Each of these procedures requires specialized equipment, and the ability to operate the equipment properly is essential to the botanist. These skills may be taught in specific courses in college or they may be learned by working in a research laboratory with a research technician familiar with the equipment.

Botany offers numerous interesting and fulfilling career opportunities. The work of a botanist is often varied as are the surroundings. Because of this great diversity in the science of botany, many people with different backgrounds, capabilities, and interests can find a rewarding career in botany.

MATCHING YOUR PERSONAL INTERESTS WITH OCCUPATIONAL AREAS IN BOTANY

Depending on your personal interests, there are a variety of botany specializations for which you are probably suited. The following are some of the specializations in which botanists work and the type of people who find these specializations interesting and satisfying.

For People Who Enjoy Research and the Outdoors

Because of the nature of their work, botanists in the following categories tend not to work, day in and day out, in the confines of research laboratories. Most work outside in areas such as the tropical rain forests, sea coasts and oceans, rivers and river beds, and fresh water bodies to identify new plant life forms. Therefore, these are occupational areas within the botany career field for people who enjoy the outdoors and who enjoy travel to remote areas of the world.

Plant taxonomists identify, name, and classify plants. (See previous section on plant taxonomists.)

Plant ecologists study the relationships between plants and the environment in which they live. (See previous section on plant ecologists.)

Aquatic botanists identify, describe, classify, and name plant life in either salt water or fresh water. (See detailed discussion later in this book.)

Paleobotanists study the evolution of fossil plants.

For People Who Enjoy Chemistry

For people who not only enjoy the study of plants but also enjoy observing, analyzing, and manipulating chemical reactions, the following occupational areas are well suited.

Plant physiologists study the functions and vital processes of plants. Some of these processes include photosynthesis and mineral nutrition. (See previous discussion on plant physiologists.)

Plant biochemists study chemical aspects of plant life processes, including the chemical products of plants (phytochemistry).

Molecular biologists study the structure and function of biological macromolecules. Molecular biology includes biochemical and molecular aspects of genetics.

Chemotaxonomists, in a subdiscipline of plant taxonomy, use the chemicals produced by plant groups to aid in their classification.

For People Who Enjoy Mathematics

For people who enjoy both mathematics and plants, there are areas of botany where numerical modeling of biological systems and applications of principles of physics contribute to the increasing body of knowledge about plant life.

Biophysicists specialize in applying the principles of physics to plant life processes.

Plant geneticists study plant heredity and variation in terms of genes and gene function in plants. (See detailed discussion later in this book.)

Systems ecologists work with the numerical aspects of plant life and its relationship to the environment. Systems ecologists use mathematical models to demonstrate concepts such as nutrient cycling in plant species.

For People Concerned about the World Food Supply

People concerned about the world food supply may be interested in the following areas:

Plant pathologists study diseases of plants. They are concerned with both the biological aspects of disease and with disease management or control. (See detailed discussion later in this book.)

Plant breeders seek to develop better types of plants. Breeding involves selecting and crossing plants with desirable traits such as disease resistance.

For People Who Enjoy Plant Structures at the Molecular Level

People who find complexity of form and design interesting may enjoy occupations that deal with the molecular level of botany. Some of these occupations include:

Bryologists—Botanists in this area study all aspects of mosses and similar plants. Bryologists are involved in the identification, classification, and ecology of these plant species.

Cytologists—Botanists in this area study plant cells and their structure, function, and life history.

Microbiologists—Microbiologists study microorganisms such as bacteria. They sometimes specialize by organisms or by branch of biology.

Mycologists—Botanists in this area study the biology of fungi. Fungi have a tremendous impact on the environment because they are crucial in recycling dead organic material. Some fungi are important producers of biological products such as vitamins and antibiotics.

Phycologists—Botanists in this area study algae. As noted previously, algae are the base of the food chain in the aquatic environments of the world. Phycologists who study algae in oceans are sometimes called marine botanists.

Pteridologists—Botanists in this area study ferns and similar plants. Pteridologists study all aspects of fern biology.

For People Interested in Artistic Uses of Plants

People who find the artistic use of plant forms and color interesting may enjoy a number of occupational fields in ornamental horticulture. Some of these fields include landscape design, interior landscape design, and floral design. They are discussed in detail later in this book.

SALARIES AND OTHER BENEFITS

Salaries for botanists depend on the level of education and specialization of the individual, the geographic area of employment and the economy of that area, and whether the botanist is employed in the private sector or by a government agency. In 1993 entry-level positions for a person with a bachelor's degree averaged $24,000 per year in private industry. Botanists with a master's degree started at $30,650 per year. Entry-level salaries for botanists with a Ph.D. are highly variable depending on the type of position (private research, government agency, or academic institution) and level of entry (research assistant, primary investigator, assistant professor, etc.). However, in 1993 an assistant professor's beginning salary averaged $37,012 per year. This figure can be misleading because average salaries varied by type, size, and geographic location of the educational institution. The average annual salary for all botanists in private industry was $35,084, and federally employed botanists averaged $41,754.

There are benefits other than salary associated with a career in botany. Many positions in botany provide rewards such as individual freedom, intellectually stimulating work associates, a pleasant work environment, varied work assignments, and an opportunity to travel. The great sense of accomplishment that comes from working in a scientific field that can bring great benefits to the world and its inhabitants is probably one of the most satisfying aspects of pursuing a career in botany.

CAREERS IN HORTICULTURE

Today there is increasing emphasis on the quality of the environment, on efficient food production for world populations, and on quality food for better health. These phenomena have created a demand for improved horticultural products and thus a new and expanded level of interest in careers related to plant cultivation and utilization.

Horticulture is one of the many divisions of the agricultural sciences. Agricultural science can include other career fields such as agronomy, forestry, soil science, plant pathology, and more. There are five major branches of horticulture: *pomology,* fruit production; *olericulture,* vegetable production; *floriculture,* flower production; *nursery culture,* nursery crop production; and *landscaping,* the design, construction, and maintenance of landscapes. Each of these divisions can be expanded further into different career fields such as the production and marketing of plant seed, nursery and greenhouse crops, and Christmas trees and turf. Other fields may include private and public gardening, wholesale and retail florist, education, research, design, advertising, photography, and food processing and storage. Even a career in a support industry such as the manufacture and sales of equipment, machinery, structures, pesticides, and other chemicals could result from an interest and training in horticulture.

In order to place the career field of horticulture within a frame of reference in the plant sciences, below is a table that defines the various fields of plant science and their relationships to one another.

Plant Science
  Agronomy and *Range Science* deal primarily with the cultivation of field
  crops and range and pasture plants.
  Forestry concerns the commercial production and utilization of timber.
    Silviculture is the practice of raising forests.

Urban Forestry is the management of trees in urban areas on larger than an individual basis.

Horticulture concerns plants that are intensively grown for food and aesthetics.

Pomology is the cultivation of perennial fruiting plants, primarily woody trees and vines.

Vegetable Crops (Olericulture) is the growing of herbaceous plants for human consumption.

Environmental Horticulture is the cultivation of plants to enhance our surroundings.

Floriculture is the production of cut flowers and potted plants.

Nursery Production (Ornamental Horticulture) is the production of primarily woody plants for landscape plantings and fruit production.

Landscape Horticulture is the care of plants in the landscape.

Arboriculture concerns the cultivation of woody plants, particularly trees.

Landscape Construction involves the installation of structural and plant materials according to a landscape plan.

Landscape Maintenance (*Gardening* or *Grounds Maintenance*) specializes in the planting and care of a wide variety of plants used in the landscape.

Turfgrass Culture concerns the growing of turf for landscape and sports use.

Landscape Architecture concerns the planning and design of outdoor space for human use and enjoyment.

Park Management concerns the total responsibility of planning, developing, and managing public and private landscaped areas, from housing developments and city parks to heavily used national parks.

From *Plants in the Landscape,* by Phillip L. Carpenter and Theodore D. Walker, 2nd edition, W. H. Freeman and Company, 1990.

WORK SETTINGS

People who make their living as horticulturists do so in a wide variety of ways and in a wide variety of work settings. Horticulturists work in laboratories, in greenhouses, orchards, vegetable fields, gardens and parks, unexplored jungles, at drawing tables, and in classrooms. They may work with a microscope, a tractor, a computer, drawing tools, or a pruning knife. Richard Adams, in a three-part series entitled "Careers in Horticulture," which appeared in the 1986 and 1987 issues of *American Horticulture,* identified "people who plant and prune, as well as people who grow and sell plants, develop budgets to buy plants, supervise planting and pruning, write and lecture about plants, and research new ways of planting and pruning" (p. 23). Such diversity of tasks and settings means that the field of horticulture can accommodate a wide range of people and interests!

EMPLOYMENT OUTLOOK

With increasing consumer interest in the environment and the efficient growth, production, and storage of food and ornamental plants, the demand for people in horticulture careers will continue to increase. While efficiency in growing plant material will definitely decrease the demand for workers who plant and grow various crops, new areas of expertise will continue to emerge.

ADVANCEMENT OPPORTUNITIES

As their experience and education levels increase, horticulturists may advance to positions with more responsibility. For example, a greenhouse technician may advance to greenhouse manager or head grower. A faculty member may advance to head of a research unit or a department head. Paths for career advancement are determined by the individual career chosen and are usually limited only by the ambitions, education level, and skills of the person pursuing the career. The nature of many of the career fields in horticulture allows horticulturists to open their own businesses.

PREPARING FOR A CAREER IN HORTICULTURE

How much education is required in the field of horticulture? Successful people in the field sometimes have less than two years of college, while others hold associate, bachelor's, master's, and even doctoral degrees. The amount of education that is required depends on the type of horticulture occupation that interests you and the level to which you wish to advance within some occupations. For example, if you wish to own and operate a florist, a specific level of education may not be required, although one or two years of college would be highly advisable. On the other hand, if you wish to teach floral design at the college level, you might need to hold at least a master's degree and possibly a doctorate in horticulture.

High School Preparation The curriculum followed by a high school student interested in a career in horticulture will depend on the specific area of horticulture of interest to the student. If he or she is not interested in a college degree, the level of advancement obtainable might be limited. Specific high school courses should include biology, math, English, general business, and computer skills. More specific courses such as plant science, horticulture, floral design, and art would be beneficial depending on the student's interest.

If the student intends to obtain a college degree in horticulture, he or she should pursue the college preparatory curriculum. Courses will include biology, chemistry, mathematics, physics, English, and foreign language. Courses in social science and humanities are required in most high school college preparatory programs. Computer courses and any courses that will help the student develop good communication skills should be considered also.

Extracurricular activities such as science clubs and science fairs will prove valuable even if the student does not intend to attend college. Summer employment in any related area also helps. This might include working for summer camps, parks, gardens, greenhouses, florist shops, plant nurseries, and landscape design and maintenance contractors. Hobbies such as gardening, photography, art, camping, and computers are also useful.

College Curriculum

In today's employment market a two-year college degree might be the minimum educational requirement for entry-level positions in some horticultural careers, but a higher level of education could be the minimum requirement for other fields. Technical positions in laboratories and in academic settings would require a four-year college degree as a minimum for entry-level positions. Advancement or other positions could require a master's degree or Ph.D. There are a few career options in horticulture for persons who do not have a college degree or technical school certificate. These employment positions within these career fields are limited, however, and advancement could be more difficult and salaries more limited than for those persons with some education and training beyond high school. For more specific information on educational requirements, refer to the section covering your particular area of interest.

Technical schools and community colleges offer certificates and degrees in horticulture. The specific courses you take will depend on the curriculum of the school and your specific area of interest. Most of these schools will require courses in biology, math, chemistry, general horticulture, and English as part of a core curriculum. Additional horticulture courses will be selected depending on your specific interest. Most of these certificate and degree requirements can be completed within two years.

Universities and colleges offering degrees in most of the areas of horticulture are land grant colleges and universities. At least one such school is located in each of the 50 states in the United States. These schools were established to teach the agricultural and technical sciences and are typically public colleges and universities.

In addition to core curriculum requirements such as English, mathematics, foreign language, chemistry, biology, social science, and humanities, students in a horticulture curriculum will take courses in general horticulture, plant physiology, entomology, plant pathology, and soil science. More specific courses are then determined by your specific career choice.

If you want to attend graduate school, you should take courses in writing, the arts, and social science. You should become proficient in using computers and a wide variety of software programs.

Summer employment, internships, and cooperative education provide invaluable work experience for students. Positions in government, college and university laboratories, agricultural and biological research stations, and private industry should be explored during the college experience.

If the curriculum permits, arrange to do an undergraduate research project under one of your professors or volunteer to assist one of your faculty members with his or her research. These kinds of experiences will not only teach you a lot about your chosen career field but can also help you to decide if a career in research and graduate school are the choices for you.

Other Personal Qualifications

Horticulturists need to be curious about the world about them and they need to be creative in solving problems. Most areas of horticulture require good writing and verbal communication skills and physical strength and stamina. An interest in nature and the outdoors is beneficial. Other personal qualifications depend on the specific career path you choose. If your area of interest is in research and teaching, an understanding of scientific method and attention to precision and detail are essential. Some areas require a flair for art and design and good interpersonal skills for working successfully with colleagues and customers.

Horticulturists are often required to understand and operate a wide variety of equipment. This equipment could be as simple as a lawn mower or pruning shears or it could be as sophisticated as growth chambers or electron microscopes. These skills may be taught in courses in college or they may be acquired by working in a laboratory or in field work.

Horticulture offers numerous interesting and rewarding career options. The work is often varied as are the work settings. The levels of education and the skills required to be successful in the field of horticulture are very broad, ranging from no college training to a Ph.D. and from the ability to operate a tractor to placing delicate flowers into an aesthetic arrangement. Because of this great diversity in horticulture, many people with different backgrounds, training, and interests can find a rewarding career in the many branches of horticulture.

SALARY AND OTHER BENEFITS

Salaries for horticulturists depend upon the level of education of the individual, the horticultural specialization, the geographic location of employment, and whether the person is employed in the private sector or by a government agency. An entry-level salary for a person without a college degree could be as low as the minimum wage, but in general average entry-level starting salaries range between $17,000 and $23,000 per year for a person with a bachelor's degree. The average horticulturist employed by the federal government earned $48,800 in 1991. Horticulturists holding a master's degree or a Ph.D. employed by colleges and universities had average annual salaries of $27,500 for an instructor-level position; $43,285 at the assistant professor level; $51,414 at the associate professor level; and $70,571 at the level of full professor. Overall, salaries will tend to be lower in more rural or economically weak areas of the country and as in other career fields, the higher your education level and the greater your skills, higher salaries and more job opportunities will be available to you.

There are more benefits to a career in horticulture than salary, however. Many of the jobs in horticulture provide rewards such as the freedom to work independently or with others who share your similar interests, a pleasant work environment indoors and outdoors, the opportunity to be creative and to solve problems, and the opportunity to travel. More obviously, the great satisfaction and sense of accomplishment that comes from working in a career field that strives to solve some of the problems associated with world hunger and environmental concerns, as well as create beautiful surroundings where people live and work are probably some of the most positive rewards for pursuing a career in horticulture.

Horticulture is a very diverse field with career opportunities for almost any person interested in the field. Jobs exist for a wide range of people from those with no formal education in horticulture to Ph.D. scientists and educators. Opportunities will continue to increase past the year 2000, especially for those with specific skills and at least a bachelor's degree level of education. The following sections describe specific careers within the field of horticulture.

POMOLOGIST

The study of the production of fruit is called *pomology,* and persons who work in this career field are called *pomologists.* Pomology is the oldest branch of horticulture in the United States. The first fruit tree nursery in America was started in 1730 in Flushing, New York by the Prince family, and the commercial production of fruit began in the mid-1880s, when shipping long distances became feasible. It should be mentioned that the term *pomology* is a misnomer. Technically, *pome* is a term used to refer to specific fruits such as apples and pears, but the term *pomology* has come to be used to refer to the study and production of any fruit.

Fruit crops include the deciduous tree fruits such as apples, pears, peaches, cherries, plums, and apricots; the evergreen tree fruits such as lemons, limes, oranges, grapefruits, bananas, and pineapples; nut tree fruits such as pecans and walnuts; and small fruits such as blueberries, blackberries, raspberries, strawberries, cranberries, and table grapes (as opposed to wine grapes). The growing and processing of grapes for wine is called *enology* and is a highly specialized type of grape production.

Fruit crops are grown throughout the United States and the rest of the world with production concentrated in areas best adapted in soils and climate to the specific crop. For example, in America citrus production is concentrated in Florida, southern California, and parts of Texas, and major areas for apple and pear production are located in the Pacific Northwest and around the Great Lakes. This type of specialization is found throughout the world.

Specific Work Performed

Pomologists propagate, plant, grow and maintain, and harvest and ship specific tree and small fruit crops. They prepare the planting areas based on the specific needs of the crop, establish the plants either by hand or with mechanical planters, water, fertilize, prune and train, harvest, package, and ship the fruit produced. Production pomologists monitor their crops for insects and diseases and apply appropriate pesticides.

Most producers specialize in a crop, but some of the larger fruit production businesses are highly diversified and produce many different crops. In many smaller fruit production operations most of the tasks described will be done by hand, but larger operations are highly mechanized and may require the operation of sophisticated equipment.

The intensity of the production process will depend on the requirements of the crop and on whether the crop is being produced for fresh consumption or for processing (canned, frozen, pickled, dried, etc.). For example, apples grown for applesauce do not need to be as blemish free, as perfectly shaped, as large, as colorful, or even as freshly picked as apples bought by the public in the grocery store.

Supervisory and management personnel in production pomology train and oversee field personnel, and they are responsible for having all work performed correctly and safely. They may coordinate all phases of the operation, set budgets, purchase equipment, and arrange for support services such as harvesting, packaging, and shipping. Some may be involved in the sales of the crop. Small businesses may require a single employee to perform several of these duties, but larger businesses typically have separate managers for the different divisions in production, sales, packing, storage, shipping, and general management. Some possible job titles you might find in production pomology might be grower, field supervisor, operations manager, field technician, and shipping supervisor, among others.

Pomologists also may be employed in research positions by private industry, government, and colleges and universities. Some pomologists are employed in teaching positions. Research pomologists study one specific fruit or a category of fruit crops to determine methods of breeding, propagating, and growing higher quality plants more efficiently and for a greater and thus more profitable yield. They test these methods both in the laboratory and in field production. Many work closely with commercial growers, addressing concerns or problems specific to that grower.

The nature of their work requires research pomologists to work with many different kinds of laboratory and field equipment. These might include sophisticated tissue culture and analysis equipment as well as more mundane equipment such as tractors, mechanical harvesters, or even hand pruners.

Pomologists involved in teaching at the college and university level often conduct research as well. Many consult with commercial growers on problems of mutual concern.

Specific Skills Required Persons employed in the field of pomology must be knowledgeable about and skilled in the propagation, growing, harvesting, and shipping of fruit crops. Most pomologists specialize in a particular fruit crop or area of fruit production such as small fruits or deciduous tree fruits, and they must have a thorough knowledge about the cultural requirements, harvesting, shipping, and handling through post production of that specialty crop.

Pomologists must be able to operate equipment associated with the production of their specific crop, such as row cultivators, mechanical harvesters, pesticide sprayers, and other such equipment. They must have a thorough knowledge of soils, irrigation practices, fertilization procedures, harvesting techniques, and pruning and training techniques. A knowledge of pest identification and control are also essential in this career field.

Anyone in a management or supervisory position in production pomology must be able to teach others how to perform the specific tasks associated with the business; thus good leadership and oral communication skills are important. A speaking knowledge of Spanish would be helpful in this field today because most laborers employed in the commercial production of fruits are Mexican or Central American. General business skills are important also.

In addition to the specific skills required of a commercial pomologist, those employed in research or academic settings must be familiar with a variety of laboratory and field equipment and techniques. A more in-depth knowledge of plant physiology, biochemistry, pathology, and genetics would be needed for success in the research branch of pomology. A good understanding of scientific methodology, statistics, and excellent oral and written communication skills are needed also.

Work Settings

Pomologists are employed in a variety of work settings such as orchards, typical farm settings, and even tropical plantation operations. Some commercial operations are small, family-owned businesses, and some can be large multinational conglomerate corporations. Commercial operations might employ only a few workers or hundreds of workers. Today there is a trend toward cooperative associations among growers to increase capitalization, storage capabilities, and marketing and shipping possibilities.

Commercial fruit production employees work outdoors, sometimes in inclement weather. Hours of work are irregular, and stressful seven-day weeks, 12-hour days are common during peak seasons. Often, work is not seasonal as in many other horticulture fields because many tasks required in fruit production, such as pruning and some pest control practices, are best performed in winter months. Because fruit production is related to the climatic requirements of the specific crop, your specific crop area may be the only factor affecting your geographic location. For instance if you are interested in growing pineapples or bananas, your work will be performed in tropical locations; and if apples are your primary interest, you will work in temperate and even cold geographic areas.

Commercial fruit production operations are located near large urban centers or in very rural settings, depending on the crop produced. Climate permitting, crops with short shelf lives or poor handling or shipping characteristics are grown close to the point of sale to reduce loss and production costs. Crops grown for processing are produced near the processing facility. Crops that store and ship easily can be produced in a wider variety of areas, again climate permitting. Thus employees in this field have many choices for the location of their work settings.

Research and teaching pomologists work in traditional research and academic settings. However, as in many other horticulture career fields, these scientists also work outdoors.

Pomology production and research usually require physical labor as a routine part of the job. This labor can vary from highly strenuous to relatively nonstrenuous. Bending, stooping, lifting, and standing for long periods are common. Accommodations can be made for persons with physical handicaps.

Employment Outlook

Food production is the largest and fastest growing area of horticulture. Increased public concern for personal health and for the environment has caused the demand for healthy nutritional food crops to increase. Fruit consumption has become more popular today because of the low-fat, low-calorie characteristics of many fruits. In addition, interest from the grower as well as the consumer in production techniques to increase yields and profitability will continue to increase the need for skilled growers and highly qualified researchers and academics to continue looking for solutions to production and other public concerns. Thus the need for highly skilled and well-trained pomologists will remain high.

Advancement Opportunities

Advancement in pomology depends on whether you are employed in a commercial or research/academic setting, a large or small operation, and on your level of education and skills. In commercial pomology, skill is very valuable; thus many people begin their careers as laborers or lower-level management or supervisory personnel. As their education levels and skills improve, they advance into more responsible positions such as grower, field supervisor, and general manager. Persons with a college-level degree may enter the profession as supervisors, graders, sales managers, and even assistant or head growers. Typically, persons in this field progress from assistant-level positions to more and more responsible management positions. Some pomologists may own their own fruit production operations.

Advancement in research and academic settings depends on your education level as well as your skills. Persons with a minimum of a bachelor's degree typically enter the field as laboratory assistants or technicians and progress as their education level increases or their skills improve, possibly to managing a laboratory or research project with other technicians under their supervision. Further advancement would require additional education or a higher degree. Pomologists in teaching positions follow the typical career path from assistant professor to associate professor to full professor, depending on degrees held and personal abilities and ambition.

Education/Training Requirements

Competition for positions in pomology production is increasing as the field becomes more and more technical and as the size of fruit production operations continues to change from the family-owned farm/orchard operation to the larger, diversified, and highly capitalized international corporation. Today, a minimum of an associate's degree in pomology is recommended for students who wish to advance in this field to career-level positions. Obviously, a bachelor's degree will make the student even more competitive for entry-level positions, and advancement should occur more quickly, depending on the person's skills. This is especially true if you are interested in employment with large commercial fruit production operations.

High school students interested in pursuing this career field should follow the college preparatory curriculum with courses in biology, chemistry, mathematics, writing and oral communication, business management, and computer skills. Working part-time in any agricultural setting provides the student exposure to techniques and equipment that will prove to be helpful.

At the community college and college and university levels students will be required to take a core curriculum plus courses in general horticulture, soils, plant pest identification and control, plant nutrition, writing and oral communication, as well as specialized courses in fruit production. To obtain a master's or doctorate degree, students will complete course work in a more specific area of pomology such as pome fruit or tropical fruit production and complete a research project and major paper under the supervision of a faculty advisor and faculty committee. Additional course work will be required in areas such as statistics.

Internships, summer employment, and cooperative education experience are helpful in gaining practical experience and training in pomology. Many opportunities will be found in small fruit production operations, commercial settings, colleges and universities, and even fruit processing plants. Although certain degrees and knowledge are required for some levels of employment in this field, like many of the other production-oriented branches of horticulture, success in pomology can depend heavily on your skills and abilities in the field rather than on any specific education level or training.

Other Personal Qualifications

A love of nature and the outdoors is essential for anyone interested in this field. In addition, pomologists need to have a concern for detail and accuracy in their work. Also, they should be observant, patient, and able to work with a wide variety of people from diverse backgrounds.

Salary and Other Benefits

As in many horticulture occupations, specific salary information for pomologists is limited and rarely reported by reliable sources. Data is based on general information about agriculture workers' salaries and is reported without reference to factors such as size of business, geographic location, and experience and education levels of employees.

Entry-level laborers should expect to make little more than minimum wage. However, persons with experience and/or an associate's degree in pomology can average $13,000 to $20,000 per year on the average. As persons advance to supervisory and management positions such as field managers and growers, their salaries will increase to $22,500 to $35,000 annually. A master's degree can increase your income up to $48,000 per year in private industry and slightly higher in government employment. Faculty teaching and doing research in pomology are paid the prevailing salaries of higher education institutions, ranging from a low of $23,700 for an instructor at the community college level to a high of $78,900 for a full professor teaching and doing research at an Ivy League university. Remember, faculty salaries vary widely by institution, location, and reputation of the individual.

Other Rewards

Pomology careers might offer the opportunity for travel. Working outdoors, working with nature, and working with a variety of people can be rewarding also. Knowing that your work is important in sustaining the well-being and lives of people throughout the world is a primary reward to pursuing a career in pomology.

FLORICULTURIST, GREENHOUSE GROWER, GREENHOUSE MANAGER

Floriculture has been an important part of the horticulture industry for a long time and it accounts for approximately one-half of the non-food horticulture industry in the United States. Floriculture is primarily a greenhouse industry except in geographic areas where the climate is mild and conducive to the growing of specific floriculture crops. The field is highly competitive with very narrow profit margins, and the industry has developed into a group of highly specialized businesses. Greenhouse floriculture is probably the most highly technical and sophisticated kind of horticulture.

The study of growing and marketing flowers and foliage plants is called floriculture, and persons working in this field are called floriculturists. Floriculturists are involved in the growing of cut flowers such as roses, snapdragons, orchids, and carnations; flowering and foliage potted plants such as chrysanthemums, poinsettias, lilies, and tropical palms, fig trees, and philodendrons; and bedding plants such as petunias, marigolds, impatiens, and geraniums. Job titles associated with this career field include propagator, grower, production manager, marketing manager, salesperson, technician, and inventory controller.

Specific Work Performed

Floriculturists are involved in the propagation and growing of cut flowers, flowering potted plants, foliage plants, and bedding plants. They plant, prune, water, and fertilize, harvest and ship these plants and monitor and manipulate the environment under which they are grown both indoors in greenhouses and outdoors in field production where the climate is favorable. Floriculturists also do research to improve the production and handling of floral and foliage crops for greater yields and improved quality. Commercial floriculturists may market their crops to retailers or to the public directly, and some floriculturists teach at the college and university level. Some floriculturists are directly involved in the daily activities of producing their crops, but others may hold supervisory and management positions. Many are responsible for the scheduling of crop production and for the mechanical operation of the greenhouse.

Growers are generally responsible for all stages of production of a single crop, a portion of a crop, or a group of crops. Growers must train and supervise workers. Large floriculture operations may employ several growers. Production managers are responsible for supervising growers, and they coordinate all production activities with efficiency and profitability in mind. Marketing managers oversee grading, handling, storage, packing, and shipment of crops. They oversee warehouse personnel who handle plants from harvest, through storage in coolers, to packaging and shipping. They supervise sales personnel and coordinate shipping. Inventory controllers are responsible for coordination of production between marketing and production departments and they typically buy all supplies and equipment.

Graders sort flower crops after harvesting into groups based on quality standards set by the industry. Propagators are responsible for increasing the number of plants in production by growing plants from seed and other techniques such as cutting and division. Technicians are most often employed in research and academic settings and assist or actively participate in the research of faculty members.

Specific Skills Required

Persons employed in the field of floriculture must be knowledgeable about and skilled in the propagation of plants and in the growing and maintenance of the plants until they are shipped to buyers. Floriculturists must be familiar with a wide variety of crops, their specific cultural requirements, and their timing, harvesting, shipping and handling through post-production. Floriculturists should be able to operate equipment commonly found in greenhouses such as pruners, sprayers, pH meters, soil sterilizers, fertilizer injectors, and heating, cooling and ventilating, watering, and humidifying equipment. They should be familiar with using a variety of light, temperature, and humidity monitoring devices, from something as basic as a thermometer to more highly sophisticated and complex equipment such as growth chambers. Many floriculture operations today are partially computerized; thus the ability to operate a computer and familiarity with basic software is helpful.

Floriculturists must be knowledgeable about soils and the unique soil mixes used in the production of floriculture crops. They must be familiar with standard potting and transplanting techniques and the specialized equipment designed to fill containers with soil mixes for high volume production. They should be able to identify typical floriculture crop pests and diseases and be familiar with proper control techniques and the operation of pest control equipment. Anyone in a management or supervisory position in the floriculture production section of the field must be skilled in all of the tasks required for production and be able to teach others how to perform these tasks as well.

General business and management skills are very important in floriculture production and a thorough understanding of greenhouse construction and the unique aspects of the operation of floriculture production environments is critical, especially for persons in management and supervisory positions. A firm grasp of the specifics of heating, cooling, and controlling other climatic factors such as humidity, and even light, is essential for the floriculturist. Today the growth of many floriculture crops is regulated by chemical and mechanical methods, and the floriculturist must be familiar with the application of these chemicals and the recommended mechanical techniques.

In addition to the specific skills required of a commercial production floriculturist, floriculturists employed in research or in academic settings must be familiar with a variety of laboratory equipment and techniques, including the use of microscopes, ranging from simple binocular equipment to electron scanning microscopes, cell and tissue culture, and staining techniques, and a wide variety of other laboratory techniques. Floriculturists employed in these areas probably should have a more in-depth knowledge of plant physiology and biochemistry than those employed in production. Knowledge of standard statistical methods, scientific methodology, and good written and oral communication skills are required also.

Work Settings

Floriculturists are employed in a variety of work settings including production greenhouses and outdoors in field production areas in geographic locations where the climate is mild or conducive to the production of certain crops. Commercial greenhouses and field production generally are limited geographically to areas of the country where the climate provides the best cultural environment such as high light conditions and warm or moderate temperatures. Typically these operations are also located near but outside of urban areas where a qualified labor pool is available, transportation costs are low, and land is available for agricultural use. States such as California, Florida, and Texas account for the majority of commercial flower production in the United States, but commercial production is found at some scale in all of the United States. Currently, the majority of cut flowers produced in the world come from Holland and Central and South America. If you are interested in traveling and working overseas, this aspect of floriculture production or research may be an ideal choice for you.

Commercial floriculture operations in the United States began as family-owned businesses and many remain so today. However, their size can range from one or two acres of production to several thousand acres. Today, many commercial floriculture businesses are subsidiaries of large conglomerate corporations.

Commercial production includes growing flowers and other crops for both wholesale and retail sales. In commercial production settings, employees may be required to work in adverse weather conditions such as extreme heat or rain, or they may work in greenhouses where temperatures can rise and fall dramatically and humidity levels are extreme. Persons involved in the postharvest handling work periodically in coolers where temperatures are low and humidity is very high.

Floriculturists are employed also in research greenhouses and laboratories by private industries, academic institutions, and a variety of government agencies.

Work environments can range from typical outdoor settings and greenhouses encompassing hundreds of acres similar to those in commercial floriculture production to confined growth chambers and sterile laboratories. Generally, floriculturists work with other people, but researchers may work alone or in contact with a limited number of technicians and other researchers. Physical demands for researchers may be similar to those required of production floriculturists; however, work environments in laboratories generally are less physically demanding and more "office-like."

Work hours in commercial production and floriculture research are often long and irregular. The demands of the crops and the research determine the work schedule, and often timing is critical for various aspects of production and for collecting research data. Also, commercial floriculturists often work in settings involved in the production of holiday crops such as poinsettias and lilies, and they must work long and stressful hours for certain peak production times such as Christmas, Valentine's Day, Easter, and the spring bedding plant season.

Employment Outlook

The employment outlook for floriculturists is good. Other than areas involved in the production of food, floriculture is the largest employer in the field of horticulture. Public concern for the environment and an overall improved quality of life has caused the demand for floriculture crops to increase steadily. As this demand continues to increase, so will the need for skilled floriculturists in commercial production settings in order to obtain high-quality and marketable products. As production demands increase, so will the need for qualified researchers and academics to look for methods to increase production yields, improve flower quality and shelf life, reduce production cost, improve profitability, and train future floriculturists.

Advancement Opportunities

Advancement in floriculture will depend on whether you are employed in a commercial or academic setting. In commercial floriculture production, experience is highly desirable; thus many people begin their careers as greenhouse or field production laborers in order to learn as many aspects of the industry as possible. They progress into positions such as assistant growers, production managers, and general managers as their skills improve or their education level rises. Persons with a college-level degree in floriculture may bypass the laborer stage and enter the profession as assistant growers, graders, or marketing personnel and advance into supervisory and management positions. Some experienced floriculturists may own their own greenhouses or field production businesses.

Advancement in research and academic settings will depend primarily on your education level. Persons with a minimum of a bachelor's degree in floriculture may enter the field as a laboratory assistant and progress to assistant technician, then technician. Advancement further will require a higher level of education. Floriculturists in college teaching follow the typical career path from instructor to assistant professor to associate professor to full professor, depending on personal ability and degrees held.

Three

Education and Training Requirements

Unless you think you will be satisfied with a laborer's position, competition for positions in floriculture will require that you have a minimum of an associate's degree for an entry-level position in this career field. A bachelor's and master's degree may be required in some industrial settings, and if research and/or teaching interest you, a master's or Ph.D. is the minimum requirement.

Students interested in pursuing a career in floriculture should follow the college preparatory curriculum in high school and take courses in biology, chemistry, mathematics, writing and oral communication, business management, and computer skills. Part-time and summer work in wholesale or retail greenhouses, nurseries and garden centers, florists, or floriculture research laboratories are advisable to gain as much experience as possible in the field.

At the community college and college and university levels students will be required to take the core horticulture curriculum with additional or elective courses in greenhouse management and crop production, herbaceous plant material identification and production, pest identification and management, business management, and writing. Persons pursuing a master's and/or doctoral degree will concentrate additional course work in a specific area of floriculture such as cut flower production, bedding plants, or foliage plants and complete a research project and a thesis paper or dissertation in the specific area of study.

Internships, summer employment, and cooperative education work experience are helpful to gain practical experience in floriculture. These types of work experience also can help the student decide which one of the many subdivisions of floriculture interests him or her most.

Other Personal Qualifications

Persons interested in floriculture as a career field should be detail oriented and accurate in their work. They must be patient, analytical, and observant. Creativity and artistic flair can be helpful. As in any horticulture career field a love of nature is essential. Most floriculture production involves physical labor, at times strenuous, requiring bending, stooping, lifting, and standing for long periods. Accommodations for persons with physical handicaps can be made.

Salary and Other Benefits

Salaries in floriculture are as variable as the tasks performed and skills required. Specific data are sketchy and rarely reported by reliable sources. Typically salaries in this career field can begin at minimum wage for uneducated and unskilled laborers, and some small production greenhouses pay little more for any entry-level position. The more education, experience, and skills a student has, the more salary he or she can expect to be offered. As responsibilities, education, and experience increase on the job, salaries will increase also.

Persons with an associate's or bachelor's degree earn $16,000 to $22,500 per year on the average. Those that advance to supervisory and management levels earn $22,000 to $30,000 per year. Persons with a master's degree employed in research can be paid up to $46,300, depending on their specific skills and background. The average salary for a floriculturist employed by the federal government is $42,750. Persons who teach at the community college and university levels average $27,650 for an assistant professor position up to $69,950 for a full professor position.

Salaries will vary geographically because of the availability of labor and the prevailing wage for a given area. Also, as in many other fields of horticulture, salaries may be based on skill and experience rather than set criteria such as education level and job title. Remember, as stated earlier, these data were inconclusive and the student would be wise to canvas employers in his or her geographic area and in his or her specific area of interest to determine the level of salary to be expected.

Other Rewards

Anyone who enjoys working with nature and in an outdoor or simulated outdoor environment will find the career field of floriculture rewarding. Satisfaction can be obtained from producing an aesthetic and marketable product from start to finish either alone or with the cooperation of a team of people. The mental challenges of working with a living product can be stimulating, and knowing that your work will bring pleasure to others is highly rewarding. If you would find travel interesting, especially to Central and South America, a floriculture career may offer that opportunity.

FLORIST, FLORAL DESIGNER

**Specific Work
Performed**

Florist and floral design careers are actually in a subdivision of floriculture. Floral designers and florists combine a knowledge of plants and flowers with design techniques to produce floral arrangements, plant gifts, decorations, and related items. Floral designers prepare corsages, bouquets, funeral pieces, dried arrangements, wreaths, and decorations for parties, weddings, and other events. Many florists manage retail shops and are involved in the daily operation of the retail business including purchasing supplies from wholesalers, taking customer orders by phone or in person, handling complaints, making deliveries, and bookkeeping. Wholesale florists purchase large quantities of flowers, plants, and floral supplies for distribution to retailers, other wholesalers in other parts of the country, or for export. They locate sources of cut flowers and supplies, prepare the flowers for storage and shipping, take orders from retailers and other wholesalers, and schedule and deliver the flowers, plants, and supplies to retail shops and others. The specific tasks performed by a floral designer and florist will depend on the number of people employed and the specific services offered by the florist, ranging from single person/owner-operated shops to large floral design/events design businesses.

Job titles for persons in this career field are numerous and might include the retail area sales clerk, designer, head designer, store manager, and buyer. In the wholesale area job titles might include receiving manager, shipping manager, salesperson, sales manager, general manager, and buyer.

Specific Skills Required   Floral designers and florists must have a good sense of design and an understanding of the concepts of form, color, harmony, scale, and other design basics. Because most of the work performed by floral designers involves the public, personal interaction skills are essential. A broad knowledge of flower varieties, their seasonal availability, and their lasting qualities is required. Because floral arrangements are prepared by hand, the floral designer must have good manual dexterity.

Work Settings   Most floral designers are employed in traditional service-oriented retail florist shops in both urban and rural settings. Some shops, particularly in urban areas, may be very large with many designers, some with specialized tasks. Other shops may be very small, a single person working as owner, manager, designer, and delivery person. These florists typically provide a range of service from customer design, potted plants, floral supplies and crafts, and delivery and wire services. There is a continuing trend for large grocery chains and other retail concerns to offer floral products and services, thus expanding the settings where a floral designer and florist could be employed. Some florists and floral designers are employed by decorator shops who custom make designs for weddings, parties, and other events. These shops generally do not work with walk-in trade and typically see customers only by appointment.

Most of the people working as floral designers and florists must stand all day and work in areas that are cool and humid to preserve flowers. Often they will move into and out of refrigerated areas. Lifting containers of flowers and supplies weighing up to 40 pounds is not uncommon, but usually materials will weigh less than 10 pounds. Adjustments can be made for persons with physical disabilities.

A 40-hour work week is common in this career field, but weekend and holiday work is common. Long work hours and working under pressure will occur around holidays such as Mother's Day and Valentine's Day, the two most popular days to send flowers.

Persons with a floral design background may also teach floral design and florist shop management at the high school, community college, and college level depending on the educational level and skills of the designer.

Florists and floral designers may also own and operate their own shops. The initial investment is high (approximately $50,000 to $75,000) and the business can be expected to turn a profit in about 7 to 10 years. Because florists depend on a perishable material for their work, the risk is high. Anyone contemplating opening his or her own shop should have a strong knowledge of the lasting qualities of flowers as well as a good background in business.

Employment Outlook   The employment outlook for floral designers and florists is good, especially in urban areas and in the suburbs of large cities. The demand for florists grows as the population increases. If the economic conditions of the geographic area served are good, florists and floral designers will be in demand. Other than specific occasions such as weddings, funerals, and certain holidays, the purchase of flowers and floral designs is closely associated with the amount of disposable income of the population served. Any downswing in the economy can cause a temporary decrease in the demand for florists. An increased awareness by the public of the aesthetic and emotional benefits of flowers and plants has resulted in a steady demand for skilled floral designers and the outlook should continue to be good.

Advancement Opportunities

Advancement in this career field will depend on the size of the business where the floral designer is employed and most importantly on the skills of the designer. Possible areas for advancement would include positions such as design supervisor, shop manager, or shop owner, or if working in an academic setting, the traditional advancement through faculty rank would be common.

Education and Training Requirements

Students in high school should include courses in biology, introductory botany, and other plant sciences. Courses related to design and art should be included also. General math, bookkeeping, and other business-related courses should be taken by students interested in floral shop management or in owning a florist's shop. Any courses that will help develop good communication and personal relationship skills should also be included in the high school student's curriculum.

If the student intends to pursue a college degree, he or she should be enrolled in a college preparatory curriculum including courses in English, foreign language, chemistry, and the social sciences and humanities.

Extracurricular activities that would prove valuable include participation in art-related activities and organizations, and science clubs and fairs. Summer jobs and internships in greenhouses, retail florist shops, and with wholesale floral suppliers are helpful also. Hobbies such as gardening, photography, and art can also help to develop valuable skills useful in pursuing a career as a floral designer and florist.

Whether or not a student intends to pursue a college or university degree, he or she should contact floral designers and florists to discuss career goals and options. A good source of information is the Society of American Florists at 1601 Duke Street, Alexandria, VA 22314.

Because a college or university degree is not required to pursue a career as a floral designer or florist, students interested in this career might consider training in adult education programs, in vocational/technical schools, and at the community college level. Courses should include design, marketing, management, and general horticulture.

Many floral designers and florists acquire their skills through on-the-job training under an experienced designer. Training would normally last approximately two years.

The specific courses you take in college will depend on the particular curriculum of the college you attend and on your specific interests. A core program would include courses in biology and general horticulture, floriculture, design, mathematics, and business management. If you intend to pursue a career teaching at the college level, you should take courses that will prepare you for graduate study such as advanced mathematics, chemistry, foreign language, social science, and humanities.

Summer jobs, internships, and cooperative education can provide important work experience. Positions would exist with retail florists, greenhouses, wholesale florists, and floral supply businesses.

Other Personal Qualifications

Floral designers and florists need a good sense of color and design as well as physical stamina and manual dexterity. Good verbal communication skills are needed to interact with suppliers and customers alike. They need to be creative, precise, and careful in their work. Florists must be aware of the necessity of meeting deadlines and public satisfaction with their work. Although some floral designers may work alone with little interaction, most benefit from good interpersonal skills for interaction with colleagues and the public. In the present job market, computer skills would be helpful for placing supply orders, inventory records, and bookkeeping.

Salary and Other Benefits

Salaries for floral designers and florists depend on the level of skill and education of the individual, the size of the business, the geographic area of employment, and the economy of the area. Entry-level floral designers and florists may be paid only minimum wage in small rural shops. As skills and creativity are proven, the pay rate will rise to $10 to $15 per hour. In 1993, the average salary for a floral designer was $23,005 per year, but a broad range of salaries was seen. Floral designers and florists in small rural businesses were paid as little as $13,000 per year, but established designers in large urban shops were paid as much as $43,950 per year.

Professionals trained in floral design and who held master's and Ph.D. degrees generally were employed in colleges and universities. These people typically had horticultural training and skills beyond floral design, in areas such as the growing of cut flowers and other floral crops and horticultural business management. Salaries were variable depending on the college or university and the academic rank held by the person. Average salaries ranged from a high of $90,900 for a full professor to a low of $22,900 for an instructor.

Other Rewards

Probably the most obvious non-monetary benefit of a career as a floral designer and florist is the opportunity to work with flowers, creating a beautiful product that will bring pleasure to the person who receives it. Most florists must also interact with the public daily, an obvious plus for anyone who does not like to work alone. Floral designers and florists are generally active; thus if you wouldn't be happy with a job that requires you to sit all day at a desk, then perhaps a career as a florist and floral designer would be rewarding for you.

OLERICULTURIST

The study of the production of vegetables is called *olericulture,* and persons who work in this career field are called *olericulturists.* Most vegetables are grown as annuals, and shifts in crops produced occur frequently and can be easily made. In the past, a large part of the vegetable production industry was diversified, but current trends are toward specialization due to rising production costs.

Vegetable crops include tomatoes, lettuce, potatoes, carrots, sweet peas, corn, onions, and squash, to name a very few examples. Note that there is no clear-cut definition for a vegetable and no clear delineation between a fruit and a vegetable. In botany, a tomato is technically a fruit, but in production a tomato is considered a vegetable. For our purposes, we will define a vegetable as any edible portion of an herbaceous plant consumed principally during the main portion of a meal. Some vegetables are valued for their botanical fruits— tomatoes and squash are examples. Other vegetable crops are valued for their edible leaves, stems, or roots. Examples include spinach, potatoes, and carrots. Most vegetables have their origins in the Old World, but crops such as potatoes, corn, peppers, and tomatoes were discovered in the Americas and exported to Europe where they became diet staples.

Vegetable crops are grown throughout the United States and the rest of the world. Large-scale production is concentrated in the United States in specific states such as California, Texas, and Florida because of numerous cultural and business factors including but not limited to labor availability, land costs, transportation, climate, and length of growing season. However, vegetable crops are produced to some extent in all of the states, indicating the flexibility of the industry.

Vegetables are produced for fresh consumption or for processing. Vegetables produced for long distance or world trade are processed to some degree due to their perishability and bulk. Fewer vegetable crops than fruit crops enter the world market, but many crops are grown for local sale or short-distance transport and do not require processing prior to sale.

The vegetable industry in the United States may be classified into the categories of market gardening and truck gardening. Market gardening refers to the production of a variety of vegetables for local and roadside markets, generally near large population centers. Market gardening is gradually disappearing in the United States because of land costs and the improvement in shipping techniques. Truck gardening refers to large-scale production of a few crops for wholesale markets and for shipping. It is dependent on climate, soils, and suitable growing season rather than market proximity. In many countries other than the United States, another classification of vegetable production is home gardening, particularly in developing countries where the home garden remains a primary source of food.

Specific Work Performed

Olericulturists propagate, plant, grow and maintain, and harvest and ship a variety of vegetable crops. They prepare the planting areas based on the specific needs of the crop, establish the plants either by hand or with mechanical planters, water, fertilize, harvest, package, and ship the vegetables produced. Producers monitor their crops for insects and diseases and they apply appropriate pesticides.

Large-scale producers specialize in a crop, particularly in geographic areas where the climate is the same throughout most of the year (southern California, for example) but many of the smaller vegetable producers are diversified, typically rotating two to three crops per year. For example, in the mid-Atlantic states, cold crops such as broccoli and/or cabbage are rotated with a warm crop such as tomatoes or corn. In many smaller vegetable production operations most of the tasks described will be done by hand, but larger operations are highly mechanized and may require the operation of sophisticated equipment.

The intensity of the production process will depend on the requirements of the crop and on whether the crop is being produced for fresh consumption or for processing (canned, frozen, pickled, dried, etc.). For example, tomatoes grown for canning do not need to be as blemish free, as perfectly shaped, as large, as colorful, or even as freshly picked as tomatoes bought by the public in the grocery store.

Supervisory and management personnel train and oversee field personnel and they are responsible for having all work performed correctly and safely. They may coordinate all phases of the operation, set budgets, purchase equipment, and arrange for support services such as harvesting, packaging, and shipping. Some may be involved in the sales of the crop. Small businesses may require a single employee to perform several of these duties, but larger businesses typically have separate managers for the different divisions in production, sales, packing, storage, shipping, and general management. Some possible job titles you might find in production olericulture are grower, field supervisor, operations manager, field technician, storage supervisor, packing and grading supervisor, shipping supervisor, salesperson, broker, and others.

Olericulturists may be employed in research positions by private industry, government, and colleges and universities. Some are employed in teaching positions. Researchers typically study one specific category of vegetables to determine methods of breeding, propagating and growing higher quality plants more efficiently and for a greater and thus more profitable yield. They test these methods both in the laboratory and in field production. Many work closely with commercial growers, addressing concerns or problems specific to that grower.

The nature of their work requires research olericulturists to work with many different kinds of laboratory and field equipment. These might include sophisticated tissue culture and analysis equipment as well as field equipment such as tractors, mechanical planters and harvesters, boom sprayers, or even hand pruners and trowels.

Olericulturists involved in teaching at the college and university level often conduct research as well. Many consult with commercial growers and government agencies on problems of mutual concern.

Specific Skills Required   Persons employed in the field of olericulture must be knowledgeable about and skilled in the propagation, growing, harvesting, and shipping of several vegetable crops. They must have a thorough knowledge of cultural requirements, harvesting, shipping, and handling through post production of the crop produced.

Vegetable producers and researchers must be able to operate a wide variety of equipment associated with the production of their specific crop, such as row cultivators, mechanical planters and harvesters, pesticide sprayers, irrigation and fertilization equipment, and other such equipment. They must have a thorough knowledge of soils, irrigation practices, fertilization procedures, harvesting techniques, and pest identification and control techniques. Like most professionals in the horticulture industry, vegetable producers and researchers work with hazardous chemicals and must be knowledgeable in the safe application of these chemicals.

A good foundation in business management is beneficial to those employed in the production area of olericulture. Anyone in a management or supervisory position in production olericulture must be able to train others to perform the specific tasks associated with the business. Thus good leadership and oral communication skills are important. A speaking knowledge of Spanish may also be beneficial today because most laborers employed in this industry are Mexican or Central American.

In addition to the specific skills required in commercial vegetable production, those employed in research or academic settings must be familiar with a variety of laboratory and field equipment and techniques. A more in-depth knowledge of plant physiology, biochemistry, pathology, and genetics would be needed for success in research. Understanding scientific methodology, a background in statistics, and excellent oral and written communication skills are needed to record, interpret, and report research data.

Work Settings   Olericulturists are employed in a variety of work settings such as greenhouses, typical farm settings, commercial and university laboratories, and many government agencies in the United States and overseas. Some commercial operations are small, family-owned businesses, and some can be large multinational conglomerate corporations. Commercial operations might employ only a few workers or thousands of workers in various locations. Today there is a trend toward cooperative associations among small-scale producers to increase capitalization and marketing and shipping capabilities.

Commercial vegetable production employees work outdoors, often in inclement weather, and in the interior environments of greenhouses. Researchers have long used greenhouses for climate-controlled scientific study. Commercial greenhouse production of vegetables is a minor but increasingly important part of the industry. In the low countries greenhouse production of tomatoes and cucumbers is a major industry, but in the United States, greenhouse production of tomatoes and lettuce for year-round consumption is a small but important and growing industry, particularly in the Midwest and Northeast.

Hours of work are long and irregular. Seven-day weeks and 12-hour days are common during peak planting and harvesting seasons. Often, work is seasonal as in many other horticulture fields, especially in areas where the growing season is limited by weather. Because of the wide variety of vegetable crops produced throughout the world, your geographic location is limited only by your interests. Commercial vegetable production operations may be located near large urban centers or in very rural settings. Climate permitting, crops with short shelf lives or poor handling or shipping characteristics are grown close to the point of sale to reduce loss and production costs. Crops that store and ship easily can be produced in a wider variety of areas, again climate permitting. Thus employees in this field have many choices for the location of their work settings, both in the United States and overseas.

Research and teaching positions in olericulture can be found in traditional research and academic settings. However, as in many other horticulture career fields, these scientists also work outdoors in the field or greenhouse.

Vegetable production and research usually require physical labor as a routine part of the job. This labor is generally strenuous, and will include bending, stooping, lifting heavy objects and equipment, and working on your feet for long periods of time. Working in extreme heat, humidity, and even cold temperatures occurs frequently, depending on the time of year and on the requirements of the crop.

Employment Outlook

Food production is the largest and fastest-growing area of horticulture. Increased public concern for personal health and for the environment has caused the demand for healthy nutritional vegetable crops to increase. The current trend toward vegetarianism appears to be more than a fad and will result in an increased demand for new and improved vegetable crops and crop production methods. In addition, interest from the grower as well as the consumer in production techniques to increase yields and profitability will continue to increase the need for skilled growers and highly qualified researchers and academics to continue looking for solutions to production problems and to address other public concerns. Thus the need for highly skilled and well-trained vegetable growers and researchers will remain high.

Advancement Opportunities

Advancement in this career field depends on whether you are employed in a commercial or research/academic setting or a large or small operation, and it depends on your level of education and skills. In commercial vegetable production, skill is very valuable; thus many people begin their careers as laborers or lower-level management or supervisory personnel. As their education levels and skills improve, they advance into more responsible positions such as grower, field supervisor, and general manager or into sales. Persons with a college degree may enter the profession as supervisors, graders, sales managers, and even assistant or head growers. Typically, persons in this field progress from assistant-level positions to more and more responsible management positions.

Advancement in research and academic settings depends on your education level as well as your skills. Persons with a minimum of a bachelor's degree typically enter the field as laboratory assistants or technicians and progress as their education level increases or their skills improve, possibly to managing a laboratory or research project with other technicians under their supervision. Further advancement would require additional education and eventually a higher degree. Olericulturists in teaching positions follow the typical career path from assistant professor to associate professor to full professor, depending on degrees held and personal abilities and ambition.

Education and Training Requirements

Competition for positions in olericulture is increasing as the field becomes more and more technical and as the size of production operations continues to change from the family-owned farm operation to the large, diversified and highly capitalized international corporation. Today, a minimum of an associate's degree in horticulture with a concentration in olericulture is recommended for students who wish to advance in this field to career-level positions. Obviously, a bachelor's degree will make the student more competitive for entry-level positions and advancement, depending on the person's skills. This is especially true if you are interested in employment with large commercial vegetable production operations.

High school students interested in pursuing this career field should follow the college preparatory curriculum with courses in biology, chemistry, mathematics, writing and oral communication, business management, and computer skills. Working part time in any agricultural setting is helpful because it exposes the student to field techniques and equipment.

At the community college and college and university levels, students will be required to take a core curriculum plus courses in general horticulture, soils, plant pest identification and control, plant nutrition, writing, and oral communication, as well as specialized courses in vegetable production. To obtain a master's or doctorate degree, students must complete course work in a specific area of olericulture such as tomato production and complete a research project and major paper under the supervision of a faculty advisor and faculty committee. Additional course work will be required in areas such as vegetable physiology, genetics and breeding, biochemistry, and statistics.

Internships, summer employment, and cooperative education experience are helpful in gaining practical experience and training in this career field. Many opportunities will be found in small vegetable production operations, commercial settings, colleges and universities, and even food processing plants. Although certain degrees and knowledge are required for some levels of employment in this field, like many of the other production-oriented branches of horticulture, success in olericulture can depend heavily on skills and abilities in the field rather than on any specific education level or training.

Other Personal Qualifications

A love of nature and the outdoors is essential for anyone interested in this field. In addition, field and research olericulturists need to have a concern for detail and accuracy in their work. Also, they should be observant, patient, and able to work with a wide variety of people from diverse backgrounds.

Salary and Other Benefits

As in many horticulture occupations such as floriculture and pomology, specific salary information is limited and rarely reported by reliable sources. Data are based on general information about agriculture workers' salaries and are reported only as averages. This information is misleading because factors such as size of business, geographic location, and experience and education levels of employees are not reported with the salary data.

Entry-level laborers should expect to make little more than minimum wage. However, persons with experience and/or an associate's degree in horticulture can average $12,000 to $18,000 per year. As persons advance to supervisory and management positions such as field managers and graders, their salaries will increase to $20,000 to $30,000 annually. A master's degree can increase income up to $48,000 per year in private industry and slightly higher in government employment. Teaching and research faculty are paid the prevailing salaries of higher education institutions, ranging from a low of $23,700 for an instructor at the community college level to a high of $78,900 for a full professor teaching and doing research at an Ivy League university. Remember, faculty salaries vary widely by institution, location, and reputation of the individual institution.

Other Rewards

Olericulture careers might offer the opportunity for travel, particularly for persons in research and academic positions. This travel may not be especially exotic, because the demand for consultants in this area is highest in third world countries. Working outdoors, working with nature, and working with a variety of people, typical in this field, can be rewarding also. Knowing that your work is important in sustaining the well-being and lives of people throughout the world is a primary reward to pursuing a career in olericulture.

LANDSCAPE CONTRACTOR, GROUNDSKEEPER, GARDENER

Landscape contractors, groundskeepers, and gardeners maintain and install landscapes at both residential and commercial sites. Landscape contractors are independent businesses contracted by the homeowner or business client for this purpose. They can range widely in size and services offered; thus career opportunities in this area are diverse. Groundskeepers and gardeners are typically staff personnel employed by businesses and homeowners directly to maintain and oversee their personal landscapes. Golf course managers and related personnel will be included in this career category although their qualifications, training, and work is more specific and unique.

Job titles in this career field are as diverse as the field itself and might include salesperson, operations manager, project manager or coordinator, construction foreman, maintenance foreman, construction and maintenance coordinator, arborist, grounds foreman, golf course superintendent, turf manager, and head gardener.

Specific Work Performed

Persons in this career field perform a very broad range of work, which will vary depending on the specific type of business or setting. In general, landscape contractors, groundskeepers, gardeners, and golf course personnel all take care of lawns, shrubs, trees, and flowers. Each may also be expected to be involved in the planning and planting of landscapes and the building and maintaining of related structures.

Landscape contractors offering maintenance services will cut, trim, water, fertilize, and rake lawns; water, fertilize, prune, and spray trees, shrubs, and flowers; and remove litter and snow. Landscape contractors involved in the installation aspect of the field buy plants from wholesale nurseries, design landscapes and gardens, grade and prepare sites for planting, and install the trees, shrubs, ground covers and turf, and related structures such as retaining walls, fences, patios, and even driveways, pools, and irrigation systems. Some landscape contractors act as general contractors, subcontracting some of the building portions of the work. Landscape contractors involved in installation often employ landscape designers or landscape architects and estimators on staff, and these businesses are sometimes referred to as design/build firms.

Some landscape contractors offer design, installation, and maintenance services to their clients, but others limit their work to one or two of these services. The size of the contracting firm and the services offered will determine the types of personnel employed and the specific tasks performed. Almost all landscape contractors employ salespersons, field foremen, and supervisors, in addition to laborers who perform the specific tasks associated with maintenance and construction.

Groundskeepers and gardeners employed on staff by businesses and individuals take care of lawns, trees, shrubs, and flowers. Specific job duties include mowing, trimming, and watering lawns and plants, pruning trees and shrubs, raking leaves, and removing litter and snow. They also apply fertilizers, herbicides, insecticides, and fungicides. They also install or oversee the installation of trees, shrubs, and other plant materials and related landscape materials. Some groundskeepers and gardeners supervise other personnel who perform the specific tasks mentioned.

Golf course personnel take care of the trees, shrubs, ground covers, and flowers on private and public golf courses. Their primary concern is the maintenance of the turf areas of the course including the rough, the fairways, and the greens areas. They must mow, water, fertilize, and otherwise intensively care for the landscape and turf areas of the course for high quality and optimal playing conditions. Most golf course workers must apply pesticides to provide a turf that meets both aesthetic and playing standards, and he or she must do so remaining aware of the environmental concerns associated with the application of chemicals, especially as intensively as needed on golf courses. Golf course personnel are also involved in planning, selecting, and planting trees, shrubs, and other plant materials and they operate and maintain irrigation systems.

Employment in these fields is sometimes seasonal, especially for field personnel. Many employers attempt to keep supervisory personnel year round by including winter services such as snow removal or holiday decorating and by emphasizing sales, planning, or other areas during the slow season.

Specific Skills Required

Given the variety of tasks performed in these fields, the specific skills required are highly varied also and they range from the ability to perform routine tasks such as pruning shrubs with a hand pruner to supervising the installation of large trees to managing the overall landscape operations of a large public facility. The skills required for persons employed in these areas might include the ability to properly operate lawn mowers, trimmers, hand pruners, forklifts, backhoes, and even larger equipment such as tree spades. Most workers in this field should be able to identify trees, shrubs, and other plants commonly found in landscapes and have an understanding of the maintenance requirements of these plants including watering, fertilizing, pruning, and pest control requirements. If the employee is involved in the installation of plants, he or she must know the proper techniques for digging, loading, transporting, and planting the trees, shrubs, and other plants. He or she should know how to prepare a site for planting and be familiar with grading, drainage, and soils. Supervisory personnel in the field must have excellent leadership and written and oral communication skills, possibly including the mastery of a foreign language. Supervisors must have the training and experience in doing the various construction or maintenance tasks they oversee, and they must be able to train workers in the procedures to be used and the equipment to be operated. Sales and business management skills are very beneficial. Anyone directly responsible for the application of pesticides will need to be licensed through a state examination process.

Work Settings

People employed in these fields generally spend a lot of their work time outdoors at the sites being planted and maintained, often in inclement weather. Supervisory, sales, and design personnel work indoors as well. Work settings include private residences, office parks and shopping centers, commercial and public sites, parks, plant nurseries, zoos, cemeteries, historic landscape sites, public and private gardens, hotels and resorts, arboreta and botanical gardens, and public and private clubs and golf courses. Work hours are long and often irregular, dictated by weather, client demands, available personnel, and work scheduled. Six-day work weeks are common, especially in spring, and physical and mental stress are common during periods of heavy work load.

Employment Outlook

This is a very large occupational field. The employment outlook remains optimistic and this continues to be a growing field of employment for a wide variety of horticulture personnel. Openings exist with small and large contracting firms, private estates, shopping centers, business and industrial parks, hotels and resorts, and golf courses. Employment with local and state governments is available in grounds maintenance and parks departments. The current public interest in the environment will continue to increase the demand for qualified landscape contractors, groundskeepers, gardeners, and golf course personnel. As demands on people's time continue to increase and property values remain high, the need for maintenance and other landscape contracting services should increase. Commercial and public sites have long understood the value of quality groundskeeping and landscape services for public relations purposes as well as the value added to the properties maintained.

Advancement Opportunities

Advancement opportunities in these fields depend on ambition, education, and experience. Many people begin working in these fields as laborers and with very small companies. As they gain experience and/or more education and training, workers can advance into positions such as foreman, supervisor, or manager, and some may advance into larger businesses or operations. Some employees may move into sales and design positions as their education and skills improve. Some workers eventually begin their own businesses. Like many of the career fields in horticulture, advancement is limited only by the individual's ability to perform the job.

Education and Training Requirements

Many groundskeepers, gardeners, and employees of landscape contractors do not have any formal education in horticulture, relying completely on prior work experience for entry into or advancement in the profession. However, as competition for positions increases, a two-year associate degree may become the minimum educational level for employment. Persons desiring to advance to supervisory and management positions should consider a bachelor's degree.

High school courses in science, mathematics, art, drafting, business, and horticulture are good preparation for this field. If the student is considering a college degree program, he or she should take the college preparatory curriculum with additional courses in business, foreign language, computer skills, chemistry, and biology as support courses. Two-year degrees preparing you for these fields will contain core courses required of all students plus specific courses in horticulture depending on the areas in which the student wishes to work. Typically students will take courses in general horticulture, plant materials, soils, chemistry, landscape maintenance and management, landscape construction, business management, and pest identification and management.

On-the-job training is common in these fields, and experience is invaluable for entry into and advancement in these career fields. However, education beyond high school will give the student greater access and opportunities to move into supervisory and management positions quickly. Certification for groundskeepers, gardeners, and golf course managers is available through state and national professional associations, but it is not necessary for entry into or success in these fields. The skills of the person performing the work is the most critical factor for success.

Summer and part-time jobs, internships, and cooperative education experience are important to help get the experience needed to be successful in these fields. Positions exist with landscape contractors, grounds departments in the public and private sectors, public and private gardens, and arboreta and botanical gardens. Some arboreta and botanical gardens even offer certificate programs that would help prepare the student for entry into these fields.

Persons desiring to enter the field of golf course management will probably need a minimum of an associate degree, and a bachelor's degree would be preferable. Certification may give the person some edge for entry into and advancement in the profession, but it is not required. Course work for the associate and bachelor's degree will concentrate on turf management, soils, irrigation, pest control practices, and grounds management and maintenance. You may be required to obtain a pesticide applicator's license. Again, any related experience prior to entering the work force is highly desirable.

Other Personal Qualifications

People who are successful in the fields of landscape contracting, groundskeeping, gardening, and golf course management are organized, artistic, and detail oriented. They have an interest in working with a diverse group of people from a variety of educational and social backgrounds. They also like to work with the public. Obviously an interest in nature is important for success in these fields.

Physical demands in this career field include the ability to lift heavy objects weighing 100 pounds or more, and unrestricted usage of the legs, arms, and hands for kneeling, reaching, stooping, and climbing. Because of the diversity of tasks in these fields, adjustments can be made for persons with physical disabilities.

Salary and Other Benefits

Data on salaries in these fields is limited and earnings will vary widely depending on the position held, size of the employer, type of employer, education, and previous experience. Laborers and entry-level employees may be paid minimum wage if unskilled or uneducated in the field. Hourly wage is not uncommon even for experienced employees and ranges from $5 to $15 per hour. Salaried employees may be paid from $13,000 to $20,000 per year, but experienced and well-educated supervisory and management level people can earn between $20,000 and $50,000 per year on average.

Other Rewards

Anyone who enjoys working outdoors and with a diverse group of people will find these career fields rewarding. The joy of working with nature and creating and maintaining aesthetic and useful outdoor environments for others to enjoy is appealing to many people. The physical labor and the mental challenges provided by this work can be highly rewarding also.

NURSERY GROWER, MANAGER

In the United States, the production of fruit crops such as apples, peaches, or even citrus fruits led to the development of the nursery industry as we know it today. Although many nurseries do produce fruit trees, the majority of nursery businesses today produce perennial ornamental plants such as shade and flowering trees, evergreen trees, deciduous and evergreen shrubs, and ground covers. A person who produces or distributes these ornamental plants is called a nursery grower or nursery manager. A person who does research or teaches in this area is referred to as an ornamental horticulturist. The American Association of Nurserymen defines the nursery industry as a major area of horticulture involved in "the production and/or distribution of plant materials, including trees, shrubs, vines and other plants that have a woody stem or stems, . . . generally used for outdoor planting by companies whose major activities are agricultural or horticultural."

The nursery industry can be divided into two categories: wholesale production nurseries and retail nurseries or garden centers. Job titles in this career field will depend on the type of nursery operation. Production nurseries typically employ plant propagators, inventory controllers, field laborers, field foremen, field supervisors, managers, salespeople, sales managers, shipping and receiving foremen, and occasionally brokers, who are independent, self-employed sales agents who locate and purchase plants from other production nurseries to fill customer orders. Retail nurseries or garden centers employ salespeople, managers, and buyers. Occasionally these nurseries will offer other services such as design and installation; thus they will employ designers and the appropriate construction personnel.

Specific Work Performed

Plant nurseries propagate, grow, and sell trees, shrubs, and other plants. The production of young fruit trees, some perennial vegetables (asparagus, for example) and some flowers, herbs, and small fruit plants are also an important part of the nursery industry. The nursery industry also produces evergreen trees for sale as Christmas trees. Nurseries may sell this plant material wholesale or retail directly to landscape contractors, to garden centers, and to landscape architects or landscape designers. Wholesale nurseries will usually limit their production to a relatively few types of plants to maximize efficiency, but some of the larger production nurseries in the United States may grow several hundred species of plants. Some nurseries will sell retail directly to the consumer. These retail nurseries often provide other related services such as design, planting, or maintenance of the plants and have a garden center as a part of their operation where they sell fertilizers, pesticides, tools, and other related materials and products. These nurseries may grow a small portion of their own plants and buy additional stock from wholesale nurseries. Some nurseries will limit their sales strictly to a mail order/catalog business. Nursery growers/managers are in charge of the overall operation of these facilities or they may be in charge of a specific portion of the nursery operation.

Work as a nursery grower/manager includes a wide variety of tasks: plant propagation through seed, grafting, cuttings, and even cell or tissue culture; greenhouse operation and management; preparation and maintenance of outdoor growing fields or container production areas; weed, disease, and insect control; plant breeding; the general monitoring, pruning, and growing of the plants; digging; loading and transporting trees, shrubs, and other plants; sales; and other business operations. In larger nurseries, these tasks are typically performed by a number of personnel, ranging from the supervisory to the general laborer level. In smaller nurseries and in many owner-operated nurseries, the nursery manager performs all of the functions with the help of a few laborers.

Working in the nursery industry may involve seasonal employment depending on the level of your employment and the geographic location of the nursery. In areas of the country where the climate is generally warm or mild, year-round employment is normal, even necessary, for a profitable operation. In areas of the country where winter weather may force outdoor work to stop or slow down for a given period of time, many laborers and some field personnel will be laid off for a brief period. Management, sales, and supervisory employees are not generally affected by the seasonal nature of outdoor nursery work.

Specific Skills Required

A thorough knowledge of plant materials and their requirements for proper growth and development into a salable product is essential. An understanding of soils, irrigation practices, pruning techniques, and pest control practices is important. Business management skills and sales ability are also valuable. Because of the wide variety of employees and customers a nursery manager or grower will encounter, good interpersonal skills are essential. Physical stamina and the strength to operate heavy machinery and equipment and to lift heavy loads may also be required.

Work Settings

Nursery managers and growers are employed in a wide variety of work settings. Some work in sterile laboratories propagating plants by cell and tissue culture. Others work in greenhouses propagating and growing containerized and other plant materials. Many nursery managers and growers work exclusively outdoors in the nursery under a wide variety of weather and physical conditions. Nursery managers and growers work with a very wide variety of other employees ranging from uneducated and inexperienced but hard-working field personnel to highly educated and experienced technicians, managers, and field employees. Thus this career field offers you the opportunity to interact with a wide variety of people with diverse backgrounds and experiences.

Employment Outlook

This is a growing field of employment at all levels. Openings exist in small local nurseries, large wholesale and retail operations, and in basic research with small and large corporations. Government employment is available at the state and national levels and state land grant universities and community colleges employ people with a background in nursery operations and management in positions such as field and laboratory technicians and nursery managers.

Advancement Opportunities

Advancement in this career field depends largely on ambition, experience, and the willingness to work long hours during certain times of the year. Many people start in this field as laborers and then work their way up through positions such as field foremen, supervisors, managers of particular operations or crops, salespeople, office workers, and others. Supervisory and management positions become available to those who acquire a broad range of knowledge and experience. Many nursery managers advance by opening their own businesses.

Education and Training Requirements

The level of education and training required for a successful career as a nursery manager or grower will vary depending on the specific level of employment and the specific position held. Today most employers prefer some formal education and training in this career field. High school courses in science, mathematics, mechanics, art, and business are good preparation for this field. Foreign language mastery is also helpful, especially Spanish. Some basic computer skills would also be expected. Even for entry-level positions, education and training beyond high school would be advantageous. Two-year degrees in nursery management, nursery crop production, and general horticulture are available from junior and community colleges and technical and vocational schools. These degrees and certificates will give the student access to entry-level management/supervisory trainee positions in small to medium-sized nursery businesses. Larger companies may require a four-year degree in nursery management, nursery crop production, or general horticulture. At either education level course work should include biology, chemistry, mathematics, writing and communication skills, and computer operations along with the other general core course requirements of the institution. Specific horticulture courses taken will depend on the specific requirements of the college or university.

Research and teaching in this career field would probably require a minimum of a master's degree or even a doctorate. Specific course work will vary with the exact nature of the degree obtained and the institution granting the degree. Some people in this field hold related degrees in areas such as chemistry, agronomy, entomology, soil science, or biology.

Because training and experience in nursery management is as important to entry into and advancement in this career field as is education, students are encouraged to seek summer employment, internships, or cooperative education work experience with nurseries, landscape contractors, and nursery industry related laboratories.

Other Personal Qualifications

Curiosity about and a love of plants is a basic requirement for anyone in this career field. Good health, average strength, manual dexterity, and a good sense of design are essential. Business management skills, sales ability, patience, and the ability to interact successfully with the public are critical in the field of nursery management.

Salary and Other Benefits

Earnings in this field vary widely depending on the size and location of the business as well as the education and training you have. Some employees in this field are paid an hourly wage ranging from $5.50 per hour to $15.00 per hour. Average salaries for nursery managers range from $12,000 to $25,000 per year. Highly educated and trained personnel in teaching or research can earn from $20,000 to $60,000 per year depending on the institution. Those working in sales often work on a commission basis, especially in retail nurseries, and their earnings range from $20,000 to $36,000 per year. Some wholesale nurseries will offer other benefits such as housing and/or a company vehicle in lieu of higher salaries.

Other Rewards

Anyone who enjoys working outdoors with living plant material and a wide variety of people will find a career in nursery management rewarding. To nurture your product over many years and then to see it used and appreciated by others is but one of the emotional rewards of nursery work. Some people will enjoy the physical labor required and the constant challenge of working with nature and the public.

Additional information about careers in nursery management and production can be obtained from the American Association of Nurserymen, Inc., 1250 I Street, N.W., Washington, DC 20005 and from individual land grant universities in each state.

LANDSCAPE DESIGNER

The practice of landscape design dates back thousands of years to ancient Egypt. The early pharaohs commissioned the design of formal enclosed gardens, leading to a highly developed knowledge of garden design that quickly spread throughout the civilized world. Many of the theories and practices of these early garden designers persist even to this day. In fact, it was those early Egyptian designers who developed the idea of using irrigation systems (canals) in order to build lush and productive gardens in the arid climate of the desert.

The general public is frequently confused or misinformed regarding the professions of garden designer, landscape designer, and landscape architect. Although these career fields are similar in some respects, they are each distinct in the actual tasks performed and in the education and training they require.

To state the obvious, garden designers design gardens. Some may have a knowledge of structural elements, and most have come into the profession out of a personal love for plants and gardens. Most are not formally trained in drafting and design, nor do they necessarily have any formal education in horticulture. Garden designers, in general, are self-taught through experience or may have had some informal training through local plant societies, arboreta, and botanical gardens.

Many garden designers do wonderful design work and are very successful in their profession due to an innate talent and years of experience working with their own gardens.

Landscape designers, as defined by the Association of Professional Landscape Designers (APLD), are "professionals who, through education, training and experience, engage for a fee in the consultation, planning, design and/or construction of exterior spaces by utilizing plant materials, and incidental paving and building materials." In comparison to garden designers, landscape designers are typically more formally educated in drafting and design techniques, horticulture, and the correct and effective usage of structural elements such as paving materials, walls, decking, and fencing.

Landscape architects, in comparison to both garden designers and landscape designers, generally work on a larger scale with municipal or commercial projects, although many do work with residential clients. Landscape architects are formally educated and trained at the college or university level or have had a minimum of ten years experience and passed a rigorous state exam. Although some landscape architects do have a strong background in horticulture, many are trained primarily in "hardscaping", that is, designing drainage systems, driveways, parking lots, retaining walls, or siting buildings or other structures.

Each of these professions contributes uniquely to the design and installation of many landscape elements from flowers to major roadways. Each has a distinct level of education and training required for success in the career field. Because of the more professional nature of the occupations of landscape designer and architect, we will explore these careers in more detail in the following pages.

Specific Work Performed

Landscape designers typically work on residential or small commercial projects. They plan and design exterior and interior spaces to satisfy the needs and desires of the client in an aesthetically pleasing way. First the designer consults with the clients to evaluate their needs and the restrictions and possibilities of the site. Landscape designers then prepare drawings for the project showing the location of plants, walkways, walls, fences, water features, and other design elements. Specific plant materials and building materials and methods are then specified. Some designers estimate the cost of the work to be performed and may act as a consultant to the client with a contractor hired to install the plan. Qualified designers may act as general contractors and oversee the entire project.

Specific Skills Required

Landscape designers must have an excellent sense of design including an understanding of the concepts of form, color, harmony, scale, and other design basics. They must be imaginative and artistic. Strong drafting skills are required for the production of accurate plan view, elevation, and detail drawings. Designers need a thorough knowledge of a wide variety of plant materials, their growth habits and requirements, as well as their proper usage to solve a variety of design problems and challenges. Landscape designers typically have a knowledge of landscape styles such as formal, informal, naturalistic, Japanese, and so on and a grasp of landscape design history. Landscape designers have a basic knowledge of soils and their properties, drainage, and building materials and methods. They have an understanding of climate and weather and the effects of rainfall, sun, temperature, and wind on plant materials and other landscape materials. Designers should be knowledgeable about zoning and building codes as well as environmental regulations. An interest in problem solving, decision making ability, and a concern for accuracy and detail is helpful for the successful landscape designer. A good understanding of mathematics and geometry, business management, and public relations is very helpful also. Because landscape designers must interact with a wide variety of people including homeowners, contractors, nurserymen, field laborers, and public officials, strong interpersonal skills are essential. Good computer skills are needed by landscape designers, including a knowledge of the more common software programs as well as some familiarity with CAD programs (computer assisted design).

Work Settings

The work settings for landscape designers can be as diverse as the tasks performed and skills required for this career choice. Landscape designers typically divide their work time between outdoor activities such as meeting with clients, evaluating sites and overseeing projects, and choosing plants and hardscape materials and indoor office duties such as meeting with clients and researching and preparing designs and drawings and a variety of support documents. Office settings will range from sparse in some small offices to elaborate in some larger offices. Some designers work alone, sometimes out of their homes; some work for landscape contractors and nurseries ranging from small to large businesses, and some designers work on staff or as consultants for architecture, landscape architecture, engineering, or land planning firms, academic institutions, or government agencies. The size and type of firm or office where the designer is employed will determine the exact nature of the work setting. However, in general, work settings for designers are comfortable and well equipped. Many landscape designers work at drafting tables, while others may work primarily with sales and client contact. Some designers draw their own plans, but others may have their drawings done by drafting personnel. Some designers work alone, but most often interaction with other designers or related staff is desirable for problem solving and the exchange of design ideas. Frequent interaction with the public and those in related and support professions is obviously necessary; thus persons preferring to work alone should reconsider this profession as a career choice.

Landscape designers also work in academic institutions and in government agencies. Designers with a master's degree and experience frequently teach design and related courses at junior and community colleges, and occasionally at the college and university level depending on the degrees earned. In this career field, experience and success in the profession can substitute for formal education. Others may be employed on college and university campuses as a part of the buildings and grounds department. Landscape designers also work for local, state, and federal agencies in areas such as urban planning, parks and recreation, environmental planning and protection, and transportation. Public and private arboreta and botanical gardens, cemeteries, and golf courses frequently employ landscape designers on staff or as consultants.

Landscape designers rarely work regular hours, the length of the work day being determined by project deadlines. Weekend work is also common, because this is often the only time clients can meet with the designer. Work is most intense during spring and summer in geographic areas where project installation can be inhibited by winter weather, but landscape designers will typically concentrate on sales for the upcoming season during the winter months.

Many people think of landscape designers as working only at designing outdoor spaces. A growing area of landscape design is the interior landscaping business. Designers involved in this area work primarily on large-scale commercial installations for shopping malls, office parks, hotels, convention and meeting centers, and other similar interior sites. The specific skills, education, and training required for an interior landscaping career are the same as those for other designers. However, the interior designer must be knowledgeable about tropical and seasonal flowering plant materials, interior irrigation systems, containers, and the appropriate climatic conditions necessary for survival of interior plants. Also, some familiarity with the impact of interior environments and public traffic on the growth of interior plants would be needed. This is a very young subdivision of the landscape design field and an ambitious and talented interior landscape designer could be very successful.

The work settings where you might find a landscape designer are highly diverse and varied in size, clientele, and the nature of the work performed. Opportunities are almost unlimited for a person interested in this career field to find a place where he or she can be happy and successful.

Employment Outlook

The employment outlook for landscape designers is strong. Because designers are able to work in a wide variety of settings ranging from self-employed independent designers to hourly or salaried consultants in large architecture and engineering firms in both the private and public sectors, the number of positions available should remain good. With increasing consumer interest in the environment and in improving the exterior and interior spaces in which we work and play, qualified landscape designers will remain in demand. Any downturn in the economy may reduce employment opportunities in firms targeting lower-end projects such as middle-income residences or small-scale commercial projects, but employment in firms targeting high-end residential and large commercial projects will remain strong. As populations continue to increase and shift from urban to suburban, the importance of usable and aesthetically pleasant exterior and interior spaces will increase. The general public is gradually accepting the fact that quality landscaping adds to the value of both residential and commercial properties, and the need for creative and qualified landscape designers will increase.

Advancement Opportunities

Advancement in the landscape design profession depends on the size and type of office in which the designer is employed. Typically, a designer may begin his or her career drafting the work of more experienced designers or architects. Advancement may result in the designer being asked to contribute design solutions or suggestions to a project or he or she may be given specific sections of the project to design or supervise. As skills increase the landscape designer may be given entire projects to oversee and eventually the size or importance of the projects may increase. This scenario may be described as advancing from draftsperson to junior designer to senior designer to project supervisor or director. Other designers may advance into sales or staff management positions, if that is the career development path they have chosen. Some designers will begin their careers self-employed and advance into staff positions with private or public employers, while others will gain experience employed by others before advancing into private practice.

Education and Training Requirements

Formal education and training in landscape design are essential for success in this career field. High school courses in biology, mathematics and geometry, chemistry, drafting, design and art, and computer skills would be highly beneficial for students considering entering this profession. Summer and after-school work with landscape contractors, nurseries, public and private gardens, grounds departments, or design firms will give the student additional training and exposure to the diverse activities associated with landscape design. Some botanical gardens such as Longwood Gardens in Pennsylvania and the New York Botanical Garden offer certificate programs in landscape design. Generally an associate or bachelor's degree is the minimum education needed for success in landscape design. In most schools these degrees would be obtained in the horticulture department. Some colleges and universities offer master's programs in landscape design. Specific course work will depend on the school attended, but generally in addition to the core requirements additional courses in design theory and principles, design history, landscape plant materials, drafting, surveying, and computer assisted design may be needed. Landscape designers also may take courses in soil science, plant pest identification and control, landscape construction, landscape maintenance, and art. Classes in technical writing, public speaking, and business management are also helpful.

Summer jobs, internships, and cooperative education can provide important work experience for students. Positions exist in private businesses, government agencies, and on college campuses.

In most landscape design programs, a senior design project is required for graduation. This type of project can help the student decide on a particular area of design he or she will like best or least and it can give the student invaluable insight into how the profession works. If the program does not require a design project for graduation, students are encouraged to volunteer to work with a faculty member on one of his or her professional projects. Students should take any possible opportunity to gain any level of experience possible, especially under the supervision of a qualified professional landscape designer.

Unlike landscape architects, landscape designers are not required to be licensed. However, there is an increased interest in the certification of landscape designers. Anyone can legally call himself or herself a landscape designer regardless of his or her actual education, training, or skill. The profession, in the interest of improving the professional image of its members as well as protecting the public from poorly trained or untrained designers has, through the Association of Professional Landscape Designers, established a certification program. A landscape designer can be certified by the APLD if he or she has a minimum of two years professional experience. The designer must also submit landscape plans that are evaluated by a committee to ensure that they meet current professional standards. Some state nursery and landscape associations offer landscape design certification through their nursery certification programs. Anyone considering landscape design as a profession should also consider the necessity of professional certification. For additional information regarding certification, contact the APLD or your local cooperative extension service.

Other Personal Qualifications

Landscape designers are typically curious and creative people who can visualize solutions to problems. They usually do not like routine and they are able to apply their education and training based on past successes and failures to solve new problems. Successful designers are rarely afraid of expressing their opinions, but they are conscious of the need for strong public relations skills. Physical stamina is needed for the long hours of design and drafting required in most practices. However, no minimum strength or other physical requirements are necessary. An appreciation of nature and the environment is basic to this profession, as is the desire to make our surroundings a more pleasant place in which to work and play.

Salary and Other Benefits

Statistics on the salaries of landscape designers are limited because of the diverse work settings and skill requirements within the profession as well as the relative youth of landscape design as a profession. In 1992 entry-level associate degree landscape designers started at an average annual salary of $16,400 as a draftsperson in larger firms in large urban areas. Those with a bachelor's degree were able to increase their salary by an additional $3,000. Many landscape designers are paid on an hourly basis ranging from $9 per hour for entry-level drafting to over $75 per hour for an experienced design professional. Compensation for self-employed landscape designers is limited only by the initiative and skill of the designer.

Salaries paid to landscape designers employed by government agencies ranged from $20,300 to $48,000 per year depending on the education and experience of the designer. Colleges and universities tended to pay on a higher scale due to the requirement for higher levels of education and training. These salaries ranged from $17,000 for an instructor to $43,300 for an associate professor and $70,500 for a full professor. It should be noted that designers holding positions of associate and full professor typically held master's or doctorate degrees and had many years of successful experience in the practice of landscape design.

Landscape design professionals are paid a broad range of salaries dependent on education level, type of work performed, size or type of employer, and geographic location. If salary is a major concern when choosing this career field, the student should seek to obtain the highest level of education possible and to develop his or her design skills to assure the compensation desired.

Other Rewards

The greatest reward for the successful landscape designer is the satisfaction of creating something that will make everyone's life more pleasant in some way. Other non-monetary rewards might include working with nature or with the variety of people who might be your clients or work associates. Although most landscape designers do not become wealthy, they share the same self-sufficiency as members of other design professions such as architects, interior planners and designers, engineers, and landscape architects.

LANDSCAPE ARCHITECT

Although a relatively new profession, landscape architecture can trace its origins back thousands of years to the early practice of garden design or landscape design in ancient Egypt. From the early Egyptians, this knowledge of garden design spread and increased throughout the Western world. In the Far East, early civilizations also developed the concept of landscape or garden design, although influenced by different principles and theories growing out of their unique culture. Out of the more ancient practices of landscape and garden design and horticulture and agriculture, the practice of landscape architecture has grown to the professional status it has today.

The American Society of Landscape Architecture (ASLA) has adopted the following definition of landscape architecture:

> Landscape architecture is the Art of design, planning or management of the land, arrangement of natural and man-made elements thereon through application of cultural and scientific knowledge, with concern for resource conservation and stewardship, to the end that the resultant environment serves a useful and enjoyable purpose.

Landscape architecture is an evolving career field. Originally landscape architects applied their skills to the design of gardens and formal estates and parks. Today professional landscape architects are also concerned with the preservation of our dwindling natural resources and the creation of beautiful and useful outdoor environments for everyone.

Specific Work Performed

Landscape architects design outdoor spaces that are functional and beautiful while taking into account the impact of the design on the natural environment. Landscape architects can work on projects ranging from small residential gardens to highly complex projects on a much larger scale such as public parks or land preservation sites. Most landscape architects specialize in a particular area of design. These areas include but are not limited to small residential design; estate design; small commercial design; the design of large-scale commercial sites such as office parks, shopping centers and housing developments, cemeteries, athletic fields and golf courses, botanical gardens and arboreta, and university campuses; public park, zoo, and recreational area design; or highway planting design. Today, some landscape architects specialize in environmental studies and preservation, land reclamation, or design of forest preserves and nature conservation sites. A few landscape architects have focused their careers on the study of landscape architecture history, writing, or teaching.

Specific work performed by landscape architects will vary depending on the area or areas of specialization chosen, but almost all practicing architects will be involved in tasks related to clients and the projects they hope to build on a specific piece of land. Exceptions might occur if the landscape architect teaches or works for a government agency.

Landscape architects meet with and interview prospective clients, inspect and analyze conditions at existing and proposed sites, write proposals and contracts, create an attractive, functional, and economically feasible design acceptable to the client using plants and a variety of structures and hardscape materials, prepare scale drawings and supporting documents for implementation of the design, solicit and review bids from contractors, and supervise the installation of the landscape plan. Landscape architects typically meet with local zoning authorities and civic committees concerned with land planning and development in their communities. In small landscape architecture practices, each landscape architect would be expected to perform most or all of these tasks, but in larger offices the landscape architect may function in a more supervisory and administrative capacity, assigning each of the aforementioned tasks to draftsmen, estimators, project coordinators, surveyors, office managers, or other assistants.

Most successful landscape architects work on several projects at the same time, and each project will probably be at a different stage. Thus landscape architects are constantly involved in the various aspects of existing contracts and at the same time analyzing new projects and contacting new potential clients.

It is important to emphasize that the specific work performed by landscape architects is more than the design and drawing of gardens. Although this is a part of the work, it is only a very small part. Landscape architects are also involved daily in the construction, engineering, political, and business details of the profession. But this diversity of work performed is one of the major aspects of this career field that makes it so appealing to those who choose landscape architecture as their profession.

Specific Skills Required   Landscape architects must be imaginative and artistic people with an appreciation for nature who can use their talents and skills to solve problems. They must possess an excellent grasp of the principles of design including form, scale, color, and other design basics. At the same time, landscape architects must be highly organized to juggle the variety of tasks performed on a daily basis. They must be well informed regarding local zoning laws, building codes, and environmental regulations. Excellent mathematical and engineering skills plus strong surveying and drafting skills are necessary for the production of accurate plan view, elevation, and detail drawings and the implementation of those drawings. A knowledge of landscape design history and the variety of landscape styles is helpful. A good understanding of plant materials and hardscape materials and their proper usage is essential. Landscape architects must have a thorough understanding of soils, drainage, and the effects of time, climate, and weather on the plants and building materials used to create outdoor spaces. Today, landscape architects need good computer skills either for preparing plans (CAD) or other documents. They need excellent written and verbal communication skills, the ability to make decisions and accept responsibility, and the ability to interact successfully with the wide variety of people they encounter from clients to contractors to politicians and civic committees. As explained earlier, every landscape architect may not need all of these specific skills to be successful in the profession. The wide variety of work in landscape architecture offers opportunities for a wide variety of people with many diverse skills.

Work Settings   The nature of landscape architecture work requires that landscape architects divide their work time between indoor and outdoor duties. Outdoors, they are involved in the site analysis and surveying aspects of the work. Indoors, they may work at drafting tables preparing drawings and other documents, or they may work in more typical office surroundings meeting with clients, discussing and presenting design ideas, or preparing proposals and contracts. Some landscape architects work alone or with a few assistants in small offices, but others may work in large architecture, engineering, or landscape architecture firms with numerous assistants and highly technical and modern equipment. In general, work settings for landscape architects are comfortable and well equipped, but the size and type of firm where the landscape architect is employed will determine the exact nature of the work setting.

Some landscape architects work for botanical gardens and arboreta. Others may work for university and college grounds departments. Historical organizations such as Colonial Williamsburg, cemeteries, resorts and theme parks, and golf courses may also have landscape architects on staff.

Landscape architects also work teaching in academic settings or for federal agencies such as the National Forest Service, the National Park Service, and the U.S. Corps of Engineers. These agencies employ landscape architects for the planning, management, and preservation of large tracts of public land. Similar positions exist at the state and local level in highway departments, parks departments, and planning departments. Some landscape architects even work overseas, especially in developing or "third-world" countries for foreign governments and private architectural and engineering firms engaged in the design and construction of new or renovated outdoor environments for those populations.

Typically, landscape architects work long and irregular hours and weekend work is common. Some landscape architects prefer working for institutions and government agencies where work hours are somewhat more regular. Work load may be somewhat seasonal, especially in smaller firms and in geographic locations where weather and climate may interfere with outdoor work. Most landscape architecture professionals concentrate on sales or long-term projects during these periods.

Employment Outlook

As public awareness and concern for the environment and for urban and regional planning continue to increase, the need for qualified landscape architects will continue to increase as well. As populations continue to increase and the need for residential, recreational, and work spaces grows, the demand for landscape architects skilled in residential design, commercial design, and planning will remain high. These factors also will continue to increase the need for employment of landscape architects by government agencies, an employment area where the number of positions has steadily increased over the past 20 years. Strong downturns in the economy or in building can have a negative effect on employment opportunities in landscape architecture, especially in small residential-oriented firms, or in larger firms highly reliant on a single type of client or work.

Advancement Opportunities

Most landscape architects begin their careers as draftsmen or apprentices for practicing landscape architects. They can expect to remain at this level for two to three years as they improve their skills. Gradually, they will be assigned responsibility for a design through more or all of its stages. As their abilities improve, they will be assigned more and more complex and difficult projects. Eventually, depending on their talent and ambition, they may become partners in the firm or advance to a higher grade in government employment. Depending on the individual landscape architect's talent, this process could take from three to ten years. Some landscape architects may advance into sales or staff management positions if that is the career path they find fulfilling.

Some landscape architects enter into private practice after gaining experience under an experienced landscape architect and when they are licensed to practice. There is an obvious high risk to private practice, thus this move should be carefully considered and a strong client base established prior to taking this step. On a more positive note, the rewards of private practice are large, and the types of projects limited only by the individual architect.

Education and Training Requirements

There are two ways to become a landscape architect. You can graduate from a college or university program accredited by the American Society of Landscape Architects (ASLA) or have practiced landscape architecture for at least ten years, and pass a state exam. Over 50 colleges and universities in the United States offer bachelor's and/or master's degrees. Most bachelor's degrees are five-year programs with partial study in the horticulture department and partial study in the school of architecture. However, some degrees are granted exclusively through the horticulture department and some are granted exclusively by the architecture school. A prospective landscape architecture student should consider these factors when choosing a landscape architecture program.

High school students should consider courses in drafting, biology, mathematics and geometry, computer skills, and art and design as a part of their core college preparatory courses. Summer and after-school work with plant nurseries, landscape contractors, engineering and survey firms, private and public parks and gardens, or even road construction and paving contractors will help prepare the student for the college curriculum in landscape architecture.

A minimum of a bachelor's degree or master's degree is required for a career in landscape architecture. Most bachelor's degrees (BLA) will take five years to complete, and most master's degrees (MLA) will take two years to complete, if you have a BLA, or three years to complete if you have a bachelor's degree in an area other than landscape architecture. A master's degree will help improve design skills with an emphasis on more complex design situations, and it greatly increases employability and starting salary.

No single curriculum is required for a degree in landscape architecture, and the courses required will vary from school to school depending on the department's and faculty's interests and expertise. In addition to the university's core curriculum, typical university and college course work leading to a degree in landscape architecture will include architecture, engineering, construction, horticulture, land planning, and landscape design. Specific courses in architecture, engineering, and construction might include training in surveying, and "hardscaping" topics such as designing driveways and parking lots, drainage systems, lighting and irrigation, siting buildings and utilities, and building berms, fences, decks, pools, patios, and other structures. Specific courses in horticulture are usually limited to basic horticulture and courses in the identification, selection, and maintenance of trees, shrubs, ground covers, and perennials. Design course work will emphasize drafting and drawing techniques specific to the profession such as lettering, perspective drawing, and plan view and elevation drawing for grading, drainage, circulation, structural, and planting plans. Additional courses in design theory and the design process, site analysis, landscape design history, and many studio courses concentrating on residential, commercial, industrial, and recreational design problems will form the bulk of the landscape architecture curriculum. Depending on the student's interests, additional courses might include urban planning, land use and land management, traffic engineering, conservation and ecology, zoning, and the legal aspects of land use. Prior to graduation most programs will include one to two courses in professional practice, teaching the student the unique aspects of landscape architecture including written and oral report presentation, marketing and contracting professional services, and a general introduction to some of the legal aspects of landscape architecture practice.

In addition to these courses directly related to landscape architecture, students should take courses in chemistry, geology, biology, written and oral communication, sociology, business management, and computer courses for drafting and word processing. Mastery of a foreign language would be helpful as larger landscape architectural firms are increasing their presence overseas.

Currently approximately 40 states require landscape architects to be licensed except those who work for government agencies. Licensing is based on passing the Landscape Architect Registration Examination (LARE), a rigorous multipart exam covering a broad range of topics such as graphics, grading, plant materials, and the history of landscape architecture. Persons taking this exam generally have a bachelor's degree and from one to four years or more of professional experience. Prior to licensing, landscape architects will generally perform all of the duties of a licensed architect but under the direct supervision of a licensed landscape architect.

Whenever possible students should seek internships or cooperative education experience in the profession or in a related field. Positions exist in private firms, government agencies, and college campuses. Because landscape architecture is a practice-oriented field, any experience would help to improve the student's knowledge, skills, and future employability and success.

Other Personal Qualifications

Landscape architects must be neat and precise and have a concern for the accuracy of their work. They are rarely satisfied with routine solutions to problems and are curious and creative. Landscape architects are rarely afraid of expressing themselves but they should be personable and always aware that strong public relations skills are necessary for a successful practice. Physical stamina is required for the long hours associated with landscape architecture practice, and minimum strength may be needed for some aspects of the outdoor work. An appreciation for the natural environment is a basic prerequisite for success in this profession, and the desire to improve the surroundings of people is essential.

Salary and Other Benefits

Statistics on the salaries paid to landscape architects are limited and compensation will depend on the educational level and skills of the architect, the types of projects on which he or she works, and the size of the firm in which he or she is employed. Landscape architects employed by government agencies typically are paid more than those in the private sector.

Entry-level bachelor's degree landscape architects were paid approximately $20,000 per year and those with a master's degree received approximately $30,000 to $30,500 annually. With experience, salaries for landscape architects will increase dependent upon the individual's skills and type and size of practice. On the average, a licensed landscape architect can make $55,000 per year or more. Landscape architects who choose an academic career are compensated at the prevailing wage for faculty rank held. In 1992, these ranged from $18,000 for an assistant professor with little teaching experience to $70,900 for a full professor with over ten years of teaching. Salaries will vary highly by the geographic location of the institution and the reputation of its landscape architecture department. Many landscape architects in academia supplement their teaching salaries with consulting and private practice. Thus, salaries for landscape architects are limited only by the skills and initiative of the individual landscape architect.

Other Rewards

Rewards other than salary associated with a career in landscape architecture include the admiration of others for the skill of your work. Working out of doors can be rewarding and the association with many other diverse people can add to the pleasure of the work. If you are interested in traveling, a landscape architecture career can provide the opportunity. Perhaps the greatest reward is the pleasure of seeing your ideas realized when projects are successfully completed and the satisfaction of knowing that your work will remain pleasing and useful to others in the future.

HORTICULTURAL THERAPIST

Specific Work Performed

Horticultural therapy is a relatively new and emerging profession, although the concept of using agriculture and horticulture for rehabilitation has been around for decades. Professionals in this area are very unique. In addition to being educated in agriculture or horticulture, they are also educated in the fields of rehabilitation, counseling, and psychology.

Horticultural therapists are trained to evaluate, rehabilitate, and train clients or patients who have mental, emotional, and/or physical disabilities using standard techniques in propagating, growing, and maintaining flowers, fruits, vegetables, trees, and shrubs or in related horticultural activities. In consultation with medical teams, horticultural therapists teach a variety of horticulture techniques for vocational rehabilitation purposes or for leisure activities.

Horticultural therapists may involve a patient in a particular phase or type of horticulture. For example, they may involve the patient in activities such as selling plants, maintaining plant materials on residential or commercial properties, or propagating plants.

Most horticulturists work with people who are also trained in the area of plant materials. However, horticultural therapists work with health care professionals. Depending on the setting in which they are employed, horticultural therapists can work with physicians, psychiatrists, counselors, social workers, and other therapists. As members of the health care team, horticultural therapists develop profiles of their clients, including an assessment of their motor skills, communication skills, and health and psychological condition. Using these data, horticultural therapists plan a therapy program appropriate for each individual. They also report patient or client progress or regression to the other members of the health care team.

The clients or patients of horticultural therapists include a wide range of individuals with mental and physical impairments. Examples of psychological or mental disabilities that might be treated by horticultural therapy are mental retardation, emotional disturbances, social maladjustments, and substance abuse.

Physical disabilities that might be treated by horticultural therapy include blindness, hearing impairment, spinal cord injury, stroke, heart attack, and cerebral palsy. Children with low self-esteem and the elderly have also benefited from working with a horticultural therapist. While horticultural therapists do work with individuals, many are more likely to work with their patients in groups.

Specific Skills Required

Horticultural therapists must possess a good knowledge of general horticulture; enjoy working with plants; and enjoy sharing that knowledge with others. They must also possess a good knowledge of the population with whom they work. For example, if a horticultural therapist works in a nursing home, he or she must possess a good knowledge of the aging process and the physical limitations and emotional needs of the people served.

In working with people with disabilities, one must have patience, understanding, and empathy. While enjoyable and rewarding, the work of the horticultural therapist can be physically and emotionally challenging. One must combine a love of plants with a strong interest in working with people.

Work Settings

Generally, horticultural therapists work in hospitals, nursing homes, schools, vocational rehabilitation centers, drug treatment centers, juvenile detention centers, and correctional institutions. Although horticultural therapy is a relatively new professional therapeutic field, there are over 1,000 programs in operation today throughout the world. Work days are usually eight hours long and somewhat under a therapist's control. Although weekend work may be expected, a work week is rarely longer than five days. Clients of horticultural therapists usually prefer routine and the therapist's work schedule generally reflects this.

Employment Outlook

Because of the growing body of knowledge about the positive relationship between the recovery process and rehabilitation therapies such as horticultural therapy, the outlook for employment is good in this area. While the first horticultural therapy programs were associated with vocational rehabilitation centers, increasingly horticultural therapists are being employed in settings such as drug rehabilitation facilities, nursing homes, and correctional institutions.

As the population ages, the need for more people in the field of horticultural therapy will increase. Other changes, such as the emphasis on assisting people with mental, physical, and emotional disabilities to work and function in the mainstream of society, will increase the demand for people who can teach horticulture skills for both avocational and vocational purposes to these diverse populations.

Advancement Opportunities

Horticultural therapists work as members of medically oriented teams. The traditional career path of being promoted to supervisor or director is rarely open to them. Advanced degrees and certification qualify horticultural therapists to teach and do research at colleges and universities. Like art, music, and dance therapists, some horticultural therapists write scholarly and technical publications to advance in the field and to enhance their professional reputations.

Education and Other Training

A bachelor's degree in horticulture is the minimum requirement for entering this field. Undergraduate programs in horticultural therapy are primarily taught in horticulture departments in schools of agriculture that are located at land grant institutions.

A typical horticultural therapy curriculum will not only stress agricultural and horticulture courses but will also include courses in psychology and rehabilitation. In addition, horticultural therapists take courses in social and behavioral sciences and specific horticultural therapy techniques and practices. Degree programs in horticultural therapy require a supervised practicum or internship.

The American Horticultural Therapy Association registers horticultural therapists. There are three categories of registration in this profession:

- Horticultural therapist technicians: These individuals have limited education and work under the close supervision of a registered horticultural therapist.
- Registered horticultural therapists: These individuals have advanced degrees in horticultural therapy and have completed at least one year of paid work experience in the field.
- Master horticultural therapists: These individuals have completed the highest level of education in horticultural therapy; have a minimum of four years of paid work experience in the field; and have demonstrated extensive educational and/or professional achievement.

Unlike professionals in other therapeutic fields, horticultural therapists are not required to be licensed in any state. However, as the profession grows, the need for certification or licensure will increase. Therefore, people engaged in this field will need to stay current in both horticulture and rehabilitation techniques.

Education and Training Programs in Horticulture Therapy

Following is a list of schools and other organizations offering training in horticultural therapy.*

Cleveland Botanical Garden
11030 East Blvd.
Cleveland, OH 44106
(Six-month internship program)

---

* Source: *Careers for Plant Lovers and Other Green Thumb Types.* Blythe Camenson (Lincolnwood, Illinois: VGM Career Horizons, 1995).

Edmonds Community College
20000 68th Ave. West
Lynnwood, WA 98036
(Two-year program in horticultural therapy)

Kansas State University
Department of Horticulture, Forestry, and Recreation
   Resources
Throckmorton Hall
Manhattan, KS 66506
(B.S. and M.S. program in Horticultural Therapy)

Kansas State University
Office of Distance Learning
Division of Continuing Education
226 College Court Building
Manhattan, KS 66506-6007
(Short-term correspondence course)

The City University of New York (in cooperation with the
   New York Botanical Garden)
250 Bedford Park Blvd., West
Bronx, NY 10468
(B.S. in horticulture with options in horticultural therapy)

Massachusetts Bay Community College
Wellesley, MA 02181
(Horticultural therapy electives)

The New York Botanical Garden
200th St. and Southern Blvd.
Bronx, NY 10458-5126
(Certificate Program—179 hours/0.5 points toward AHTA
   Professional Registration)

Rockland Community College
Suffern, NY 10901
(Horticultural therapy electives)

Temple University
Department of Landscape Architecture and Horticulture
Ambler, PA 19002
(Horticultural therapy electives)

Tennessee Technological University
School of Agriculture
Box 5034
Cookerville, TN 38505
(Horticultural therapy electives)

Texas A & M University
Department of Horticulture
College Station, TX 77843-2133
(B.S. in horticulture with options in horticultural therapy)

Tulsa Junior College
Northeast Campus
Department of Science and Engineering
3727 East Apache
Tulsa, OK 74115
(Horticultural therapy electives)

University of Massachusetts
Department of Plant and Soil Science
Durfee Conservatory, French Hall
Amherst, MA 01002
(Horticultural therapy electives)

University of Rhode Island
Department of Plant Science
Kingston, RI 02881
(B.S. in horticulture with options in horticultural therapy)

Virginia Polytechnic Institute and State University
Department of Horticulture
Blacksburg, VA 24061
(Horticultural therapy electives)

Other Personal Qualifications

Horticultural therapists must enjoy working with a wide variety of people. They must be patient, well organized, reliable, and empathetic to the problems and needs of others. Physical stamina is necessary but there are no specific requirements for strength or mobility. Horticultural therapists may work under conditions of mental and emotional stress and should be able to cope with stressful situations and difficult clients.

Salary and Other Benefits

Salaries of horticultural therapists vary according to the employment setting and the educational preparation. According to the U.S. Office of Personnel Management, rehabilitation therapy assistants, those with minimal educational preparation, earn an estimated average salary of $21,500 per year.

By contrast, rehabilitation therapists at the baccalaureate degree level can earn an average salary of approximately $35,000. Experienced horticultural therapists can make as much as $50,000 per year.

Beginning salaries for rehabilitation therapists working in hospitals average $32,000. In the federal government, where many therapy-related professionals are employed by the Veterans Administration, the average salary is approximately $31,000 per year. Supervisory salaries average about $42,000 per year.

Other Rewards

Knowing that you are improving the quality of someone's life is a major secondary benefit to a career in horticultural therapy. Work settings are typically pleasant and clients generally show appreciation to the therapists for skills learned.

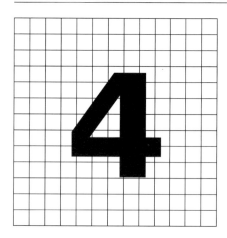

CAREER FIELDS RELATED TO BOTANY AND HORTICULTURE

The career fields of botany and horticulture are so broad and diverse that there are specific career options that cannot be easily classified under a particular heading. Though not specific to either career field, the following divisions of botany and horticulture should also be considered.

PLANT PATHOLOGIST

Plant pathologists are botanists who study diseases that afflict plants. Plant pathologists conduct studies on the nature, cause, and control of plant diseases as well as the decay of plant products. Often their studies compare healthy plants with diseased specimens to determine the agent responsible for the disease. They are concerned with the spread and intensity of disease under different conditions of soil, climate, and geography. They often predict outbreaks of plant disease and try to determine which kinds of plants and insects will transmit the disease. Plant pathologists are concerned with both the biological aspects of diseases and with disease management and control.

Plant pathologists attempt to isolate disease-causing agents and study the habits and life cycles of various plants. Many times the goal of the plant pathologist is to destroy or control the agent that is causing the disease. Dutch Elm Disease, for example, is a disease many plant pathologists have studied in an attempt to identify the agent and to control its spread. Unfortunately many elm trees had to be destroyed in order to halt the spread of the disease. Plant pathologists also test possible disease control measures under experimental conditions in their laboratories and research fields. They study the comparative effectiveness of different treatments and make recommendations based on practicality, economy, and safety.

PLANT GENETICIST, PLANT BREEDER

Plant geneticists are botanists who study plant heredity and plant variation. Plant geneticists perform experiments to identify the biological laws and mechanisms, as well as the environmental factors, that are central to the origin, transmission, and development of inherited traits in plants.

In addition, plant geneticists analyze the determinants of specific inherited traits. These traits may include such factors as color differences, plant size, and disease resistance. Knowledge gained through this research can be used to improve or understand the relationships of heredity, maturity, fertility, or other factors.

Botanists who work in the area of plant genetics devise methods for altering or producing new traits in plant species. They make use of chemicals, temperature, light, or other means to study the plant/environment relationship.

Plant geneticists may specialize in particular branches of genetics such as molecular or population genetics. Probably the best-known plant geneticist in the area of molecular biology is Barbara McClintock. Dr. McClintock won the Nobel Prize in 1983 for her pioneering work in corn genetics. As a plant molecular biologist, Dr. McClintock's work involved the transposable elements of genes (referred to as "jumping genes") and the plant/environment interaction. These discoveries led to the development of a stronger species that can resist disease, pests, and climate changes, and they have contributed to the development of a more durable food crop for distribution and consumption. The techniques and principles developed by Dr. McClintock continue to be applied to other species and types of studies by plant scientists.

Recently the first genetically altered food crop, a tomato, reached store shelves. This tomato has a much longer shelf life than other tomatoes without loss of flavor. This has signaled the beginning of a new career field for botanists. Plant biotechnology involves modifying or inserting desirable genes into plants and having those genes expressed. Biotechnology uses this genetic manipulation of biological organisms to reproduce useful products.

In the related occupation of *plant breeder,* the research of plant geneticists is used to develop and improve varieties of crops. Plant breeders improve specific characteristics, such as yield, size, quality, maturity, and resistance to frost, drought, disease, and insect pests. They accomplish this by utilizing both the principles of plant genetics and the knowledge of plant growth.

Plant breeders develop varieties and select the most desirable plants for continued reproduction using methods such as inbreeding, crossbreeding, backcrossing, outcrossing, mutation, or interspecific hybridization and selection. They select offspring of plants that have the desired characteristics and they continue the breeding and selection process until the desired characteristics are consistently expressed by the majority of new plants.

Through their work, plant geneticists and plant breeders have advanced the field of biotechnology in the area of botany. According to the Botanical Society of America, plant geneticists have opened a new career field for botanists: "In fact, biotechnology uses biological organisms to reproduce useful products. Most people today have a narrower view of biotechnology as the genetic modification of living organisms to produce useful products. Plant biotechnology involves inserting desirable genes into plants and having those genes expressed." As a result, the research of plant geneticists has the potential to improve domestic plants.

AQUATIC BOTANIST

Aquatic botanists study plants that live in water. Aquatic botanists who identify, describe, classify and name plant life in fresh water are also known as *limnologists.* Aquatic botanists who study salt-water organisms are known as *phycologists* because they study algae. Their work is of significance because algae is near the base of the food chain in the aquatic environments of the world. Botanists who study salt-water plant life are also called *marine botanists.*

AGRICULTURAL ENGINEER

In VGM's *Careers in Engineering* (Garner, 1994), agricultural engineering is described as one of the basic engineering disciplines with the closest relationship to horticulture and botany of any engineering discipline. Agricultural engineers ensure tomorrow's food production.

Agricultural engineers work with horticulturists to design equipment and processes to plant, harvest, produce, and distribute food stuffs. As agricultural engineers develop new tools it becomes easier and more practical to produce, process, and distribute food. Using scientific principles, agricultural engineers design systems and equipment to manage the various resources that horticulturists and botanists need to produce food. These resources include soil, water, air energy, and engineering materials. Agricultural engineers apply their skills across the entire food production chain, from the protection of natural resources to the preservation of food products.

Some agricultural engineers are employed in the increasingly technological agricultural industry. Still others run experimental farm stations and research laboratories, usually at land grant universities, that benefit the agricultural community and industry. Other agricultural engineers work as consultants for projects in another country.

With society's increasing concern about world hunger, the agricultural engineer plays an important role on the team of botanists and horticulturists who plan and implement new technologies for food production.

Unlike the preparation of horticulturists and botanists, agricultural engineers concentrate their studies in mechanical, civil, electrical, chemical, and other types of engineering, in addition to biology and horticulture training. Universities that specialize in preparing agricultural engineers are located at most of the land grant universities, where courses in both engineering and agriculture are offered.

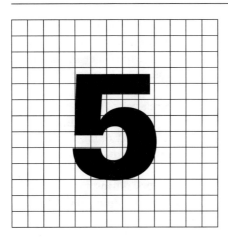

TEACHING OPPORTUNITIES FOR BOTANISTS AND HORTICULTURISTS

People who pursue undergraduate degrees in botany and horticulture have the career option of teaching young people about their discipline in private and public schools, primarily at the secondary level. To teach at this level, you must not only complete an undergraduate degree in botany or horticulture, you must also become certified in the state in which you want to teach.

Likewise people who pursue a master's degree in botany and horticulture will find that junior and community college teaching may be a viable career option. Teaching at this level involves preparing students to go directly into industry or to pursue a B.S. degree at a four-year college or university. This means that people who teach at this level must have a good knowledge of industry, as well as higher education requirements beyond the associate degree level.

People who pursue a Ph.D. degree and who are involved in research in botany and horticulture will find teaching at four-year universities or two-year colleges a viable career option. In addition to preparing future generations of botanists and horticulturists, college faculty enjoy learning and they actively contribute to the body of knowledge in their field.

TEACHING AT THE SECONDARY LEVEL

Elementary teachers introduce young students to very basic facts about plants, flowers, and trees through science courses and special projects. However, it is the secondary teachers in biology and vocational education courses who introduce high school students to more in-depth knowledge about plant material. They cover such concepts as photosynthesis, propagation, and plant nutrition.

Teachers at this level can expect to supervise laboratories as well as teaching classes in their subject area. Vocational education teachers, who teach horticultural skills, may often supervise student work experience in nurseries, greenhouses, and commercial business such as florists and landscape design firms.

In addition to classroom responsibilities, secondary teachers, regardless of the subject matter they teach, are responsible for advising students and meeting with parents. Many teachers also advise student organizations and participate in other aspects of school life, such as serving on school and PTA committees.

While the public may think that teachers only work during the hours that students are in school, that is far from true. Course preparation, student evaluation, and school service make teaching at least a 40-hour-a-week job, often more hours than that on the teacher's own time. While there are some year-round school districts, the majority of teachers work ten months a year and have a two-month, non-paid vacation during the summer. During the summer, many teachers take college courses in order to stay current in their fields. Others teach in summer school.

In most states, teachers serve a probationary period. This is usually three years. After the probationary period, teachers are considered to be "tenured". This means that they cannot be fired without just cause and due process.

Teaching at the secondary level requires a minimum of a bachelor's degree in the appropriate area and teacher certification in the state in which the person plans to teach. While certification requirements may differ from state to state, there are some common elements in the certification process for all teachers. According to the *Occupational Outlook Handbook,* published by VGM, "Certification is generally for one or several related subjects. Usually certification is granted by the State board of education or a certification advisory committee. . . . A bachelor's degree and completion of an approved teacher training program with a prescribed number of subject and education credits and supervised practice teaching in a secondary school [are required]."

There are other methods for becoming certified in some states. These approaches include:

- Beginning to teach immediately upon graduation under a provisional certification, after working under the close supervision of experienced teachers for one or two years while taking education courses;
- Taking only those courses that are lacking for certification. When the courses are completed, certification is granted;
- Entering a program that grants a master's degree in education, as well as teacher certification;
- Obtaining an emergency certification in a school district that is unable to hire enough teachers with regular certification.

Earnings

The following table shows the average annual salaries of public school teachers in each state.

Average Annual Salaries of Public School Teachers

State	Average Salary	State	Average Salary
Alabama	$26,954	Nebraska	$27,231
Alaska	44,725[1]	Nevada	33,175
Arizona	31,892[1]	New Hampshire	32,445
Arkansas	26,569	New Jersey	41,381
California	41,811	New Mexico	26,653
Colorado	32,926[1]	New York	44,200
Connecticut	47,300	North Carolina	29,334
Delaware	34,548	North Dakota	24,145
Florida	31,119	Ohio	34,359
Georgia	29,680	Oklahoma	25,721
Hawaii	34,528	Oregon	33,656[1]
Idaho	26,759	Pennsylvania	38,540
Illinois	36,623	Rhode Island	36,047
Indiana	33,755[1]	South Carolina	28,209
Iowa	29,196	South Dakota	23,300
Kansas	30,808	Tennessee	28,726
Kentucky	30,880	Texas	29,041
Louisiana	27,037[1]	Utah	26,524
Maine	29,672[1]	Vermont	33,420
Maryland	38,843	Virginia	31,921
Massachusetts	38,066[1]	Washington	34,880
Michigan	40,251	Washington, D.C.	41,256
Minnesota	34,782	West Virginia	27,298
Mississippi	24,428	Wisconsin	33,873
Missouri	28,880	Wyoming	29,000
Montana	27,513		

[1]Estimate

Source: National Education Association (1990-1991).

Career Outlook

There is an increasing demand for teachers with a scientific background. Therefore, people with undergraduate degrees in botany, or biology with an emphasis in botany, will find job prospects to be very good provided they have the necessary certification.

Not every secondary school will offer vocational education programs in horticulture, thus limiting the opportunity for people with horticulture degrees to teach their specific discipline. However, if they have completed a sufficient number of biology or chemistry courses in their undergraduate preparation, they may be able to become certified to teach in these disciplines within the state system where they seek certification.

However, people with undergraduate degrees in horticulture are not limited to teaching only within the public school setting. Opportunities exist to teach horticulture skills and knowledge in nontraditional settings such as vocational rehabilitation centers or correctional institutions. Therefore, the opportunities to teach horticulture are varied.

Advancement Possibilities

Secondary teachers can advance through a system of promotions to become "master teachers". Some also move into counseling or administrative positions. Administrative positions within a school can include department chair, assistant or vice principal, and principal. Administrative positions within the school system may include supervisor of curriculum or instruction at the local or state level.

More Information on Teaching at the Secondary Level

To learn more about teaching opportunities and certification requirements in your area of interest, write to the professional associations related to the specific discipline you plan to teach as well as to the local school system and the state department of education where you plan to teach. Other resources on teaching careers include:

> National Education Association
> 1201 16th Street, NW
> Washington, DC 20036

> American Federation of Teachers
> 555 New Jersey Avenue, NW
> Washington, DC 20001

In addition, Recruiting New Teachers, Inc. provides a toll-free number (1-800-45-TEACH). It offers guidance on how to become a teacher and sends the names of prospective teachers to a network of school districts and teachers' colleges.

TEACHING AT THE COLLEGE AND UNIVERSITY LEVEL

Botany faculty teach at a wide variety of colleges and universities. Of those who teach at four-year institutions, some teach at major research institutions. Others teach at land grant colleges and universities. Still others teach at liberal arts colleges and universities. In contrast, horticulture faculty primarily teach at land grant colleges and universities as well as two-year and community colleges.

Some faculty pursue research as the major emphasis in their faculty role. However, while research funding and publications may be of paramount importance in nearly every college and university, poor teaching is viewed negatively even at institutions that place primary emphasis on research.

Because the preparation of future botanists and horticulturists is vital to the profession and to the nation, teaching is at the center of any academic career. However, most four-year colleges and universities tend to emphasize the importance of faculty research. Two-year and community colleges place primary emphasis on teaching. However, in the area of horticulture, both two-year and four-year institutions place importance on the relationship between the faculty and farmers and/or the agricultural industry.

Faculty employed at land grant universities can have "split appointments". This means that they can have a primary appointment in the areas of teaching, research, or extension. If their appointment is primarily in the area of teaching, they will be evaluated on their teaching innovation and abilities. Those who have a primary appointment in research will be evaluated on the quality and significance of their basic or applied research in their field. In the area of extension, faculty are evaluated on their service to industry and the public.

An important aspect of an academic career for all faculty is interaction with students, whether undergraduate or graduate students, and their academic colleagues. Faculty enjoy a special fellowship and camaraderie among themselves and their students. This interaction usually transcends the campus to include faculty, students, and professionals around the country and the world. These scholarly interactions have meant that faculty have almost always had a lead role in the pursuit of scientific and technological advances in botany and horticulture.

Working hand and hand with the government and industry, faculty research has advanced the knowledge base in botany and horticulture. The rapidly increasing interest in the knowledge and skills essential to grow and market horticultural crops is related to the world's concern for environmental quality and health. Therefore, the role of botany or horticulture faculty members has become even more significant.

Faculty are at the center of the transfer of scientific knowledge for industrial and consumer usage. Through scholarly research, innovative teaching, and major contributions to the literature of their field, faculty are instrumental in impacting the overall quality of life. Therefore, if you enjoy the process of learning and growing intellectually, if you like research and laboratory work, an academic career in the area of botany or horticulture may be a very appropriate career path.

Where to Begin

Interestingly, an academic career does not begin with your first faculty appointment. It actually begins when you start to think about going to graduate school. In some instances, you may find that you are being advised to obtain experience in the field before pursuing a graduate degree. This advice is not contradictory. It is quite sound in today's academic environment. Experience, as an undergraduate intern, as a co-op student, or as a full-time employee, will provide an excellent foundation for an academic career. There are two major benefits to this approach.

First, as a future faculty member your experience will provide a better understanding of the issues that your students will encounter in the field. This is extremely important because most students will pursue career paths in business, industry, or government. Your understanding of the issues and constraints they will face will be of significant importance in how you advise and teach them. Second, practical experience will help you focus more clearly on the objectives of your graduate study. It may even help you determine the subspecialties you wish to pursue and the best faculty person with whom to study in that area.

Upon entering graduate school, your appointment as a graduate research assistant or a graduate teaching assistant can form the basis for a successful academic career. Prestigious fellowship awards to support your graduate education may also help you gain a faculty position after you complete your graduate studies.

Today, it is almost a necessity to have post-doctoral research experience before receiving an appointment on a college or university faculty. A track record of successful research and scholarly publications is an important credential when competing for faculty positions. Your post-doctoral time, usually at a university different from the one awarding your Ph.D., provides you with additional time to build your scholarly reputation in the field.

Appointment as an *assistant professor* in a "tenure track" position is the point at which you become a member of the faculty. Tenure track assistant professors have up to seven years to become tenured, but in reality they have six years to build their credentials for tenure. If you are not granted tenure and promoted to the level of *associate professor,* you have the seventh year to seek positions outside of the academy or at another institution.

In general, *associate professors* remain active in their teaching, research, and/or service to the profession. By doing so, they may qualify for promotion to *full professor.* This is the highest academic appointment that one can receive.

Working Conditions

The working conditions in colleges and universities differ widely. Each campus has its own unique personality and style. An important determinant of these differences is the mission of the overall institution and the focus of the department. The differences also are attributable to the individuals who make up the faculty.

Every college and university catalog clearly states the institution's mission. Prospective faculty members should read those statements carefully. In addition, if the department has its own mission statement, that should be examined as well. These statements of purpose will tell you what activities the institution values and consequently which activities it will reward.

Some institutions emphasize faculty research and publications. Faculty in these settings are expected to set a research agenda for themselves, secure outside funding from industry, foundations, and/or government agencies, publish their results, sponsor colloquia and symposiums, and become recognized experts in their specific area of engineering. Schools that are ranked among the top 25 graduate programs in the country tend to fall into this category.

Other schools emphasize teaching. In these unique settings innovative teaching techniques and publication of textbooks are expected. The faculty/student ratio is an important consideration because the academic development of the students is critical to the mission and goals of this type of institution. This is not to say that teaching is undervalued in research-oriented universities. All institutions of higher learning strive to provide a high level of teaching. However, at institutions that focus exclusively on teaching, faculty members are held more accountable for their expertise in the classroom, as opposed to their research expertise.

In addition to teaching and research-oriented institutions, there are also institutions whose mission it is to provide extension services to farmers, industry, and home owners. Land grant universities are excellent examples of institutions that are committed to professional practice in the various specialties of the agricultural industry. These institutions provide a valuable service to business, industry, and government.

Excellent teaching, research productivity, and extension service do not automatically mean that faculty can continue to teach and do research indefinitely. Most faculty have to be reviewed by a rigorous tenure process at the end of their fifth year. While some institutions do not have, or have abolished, tenure, the majority still maintain this process of peer review to assure the quality of the faculty.

Earnings

It is important to know that faculty salaries at both two-year and four-year institutions reflect nine- or ten-month contracts, not twelve months of income. In order to assure a year-round salary, faculty must budget to live on the base salary on a twelve-month basis, or they must secure summer teaching and/or research funding to increase the base salary. However, at some institutions faculty may only cover a portion of their summer salary by these means.

As faculty develop national, state, and local reputations in their field, they can augment their salaries with consulting fees. Business, industry, and government seek the expertise of college and university faculty who have established their expertise in a specific area. Royalties from books, software, and/or patents also contribute to a faculty member's overall income.

In comparing faculty base salaries in botany and horticulture, *The Chronicle of Higher Education* annually publishes an extensive breakdown of all faculty salaries by rank, discipline, and institution. This is a good resource for obtaining up-to-date salary information for college and university faculty across the country.

Career Outlook

Despite the earlier predictions of shortages in qualified faculty, the career outlook in academia is not bright at the present time. Cutbacks in federal and state funding of higher education and research; declining population among college age students; and instructional costs are some of the factors operating to keep the number of new faculty positions low. When positions do become available, competition is stiff. Candidates need to be flexible and well prepared. They must also have strong motivation and a high energy level because it is also increasingly difficult to get tenured at leading universities.

Advancement Possibilities

At most colleges and universities there are standard advancement opportunities. Faculty are promoted from assistant professor to associate professor to full professor. In addition, there are administrative positions which faculty members can hold. In some cases these positions are viewed as promotions, and in other cases these positions are viewed as temporary appointments. After holding one of these positions as a temporary appointment, the faculty returns to his or her academic department to resume an academic career of teaching, research, and/or extension. The following list demonstrates the range of opportunities that exist for botany and horticulture faculty: post-doctoral fellow; adjunct professor; research associate; lecturer; assistant professor, tenure track or non-tenure track; associate professor, tenure track or non-tenure track; full professor, usually only tenured but very occasionally non-tenured; program director; department chairperson; assistant dean; associate dean; dean; vice president of research, graduate studies, or extension.

Professional Associations

Almost every professional society in botany and horticulture represents the interest of its faculty members as well as its members in business, industry, and government. See the Appendix for a complete listing of professional societies and associations.

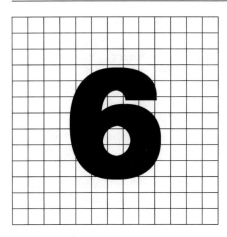

GOVERNMENT OPPORTUNITIES FOR BOTANISTS AND HORTICULTURISTS

Many botanists and horticulturists are employed by government agencies at all levels: federal, state, or local. However, in recent years, opportunities for government employment have declined significantly. Nonetheless, government careers in such agencies as departments of agriculture, parks and recreation, and environmental protection offer excellent opportunities for challenging and fulfilling experiences. In addition, they offer a wide variety of career paths for botanists and horticulturists.

FEDERAL GOVERNMENT JOBS

According to the U.S. Office of Personnel Management (OPM), the federal government hires between 200,000 and 300,000 people annually. A good number of these are botanists and horticulturists. They are employed in such agencies as the National Park Service, the U.S. Department of Agriculture, and the Environmental Protection Agency. The work that they do and the background and experience required vary from discipline to discipline and agency to agency.

An interesting aspect of career opportunities in government agencies is that they often involve interaction with private sector companies, college and universities, and foreign governments. This variety of interaction and opportunity provides numerous career paths.

According to Petras and Petras, government careers can be very satisfying. Compensation of botanists and horticulturists employed by government agencies tends to be competitive with salaries found in the private sector. In addition, the fringe benefits associated with government employment are good.

The work in federal agencies is often on the "cutting edge", particularly in terms of research and development. Consider the National Science Foundation (NSF). The NSF is an independent agency of the federal government that is committed to promoting and supporting research and education programs in science and engineering. To accomplish its goals, the NSF establishes cooperative agreements with universities, consortia, and nonprofit research organizations in order to fund major research projects in order to advance knowledge in a particular field. For instance, research may be funded in the area of plant genetics or hydroponics in order to develop stronger varieties of food crops to feed people in third world countries.

There are other federal agencies that hire significantly more botanists and horticulturists than the NSF. They are the National Park Service and the U.S. Department of Agriculture.

The National Park Service is part of the U.S. Department of the Interior. The Interior Department is the largest conservation agency in this country. Because the Department is responsible for the preservation of public lands and natural resources, the National Park Service is a vital part of the agency's work.

The Park Service offers opportunities to plan, design, implement, and preserve such areas as historical and botanical gardens, national parks, and the grounds of historic monuments. This means that the work of the Park Service is carried out in large and small communities throughout the United States.

In *The Almanac of American Government Jobs and Careers,* the Department of Agriculture is described as having responsibility for "implementing agricultural policies designed to maintain and improve farm income; develop and expand markets abroad for U.S. agricultural products; prevent poverty, hunger, and malnutrition; enhance the environment and productive capacity through soil, water, forest, and natural resources conservation; promote rural development, credit, and conservation programs; conduct research; and maintain the quality and safeguard the daily food supply through inspection and grading services" (p. 33). The divisions of the Department of Agriculture reflect this work. The divisions are Small Community and Rural Development; Marketing and Inspection Services; Food and Consumer Services; International Affairs and Commodity Programs; Science and Education; Natural Resources and Environment; and Economics. The diversity of the work performed by the U.S. Department of Agriculture provides a wide variety of career opportunities for botanists and horticulturists.

The work of the Department of Agriculture is not just carried out in Washington, D.C. Throughout the United States, there are numerous county and district offices and agencies of the Department of Agriculture. In addition, there are offices in foreign countries. In fact, nearly 90 percent of the work of the Department of Agriculture is performed outside of Washington, D.C. While the Department of Agriculture hires a wide variety of professionals to accomplish its goals, many of those professionals are horticulturists and botanists.

The salary rates for botanists and horticulturists employed by the federal government are as follows:

GS Level	Average	Minimum (Step 1)	Maximum (Step 10)
GS-5	$18,699	$16,305	$21,201
GS-7	23,005	20,195	26,252
GS-9	27,793	24,705	32,121
GS-11	33,812	29,891	38,855
GS-12	40,801	35,825	46,571
GS-13	49,003	42,601	55,381
GS-14	58,363	50,342	65,444
GS-15	70,316	59,216	76,982

The Future Outlook for Federal Employment

Despite budget deficits and a general slowdown in government hiring, the Office of Management and Budget predicts that the demand for botanists and horticulturists will be higher than average because of new areas of biotechnology and increased concern for environmental issues. While most horticulturists' and botanists' jobs are in the executive branch of the federal government (i.e., Department of Agriculture, Department of Interior), opportunities also exist in the legislative branch (for example, U.S. Botanical Garden) and in independent agencies and government corporations such as the Environmental Protection Agency or the National Science Foundation.

STATE AND LOCAL GOVERNMENT JOBS

There are many opportunities for botanists and horticulturists to work at the state, local, and federal levels. Traditionally, state departments of agriculture have been the principal employers of horticulturists and botanists.

Usually competition for state and local government jobs is not as stiff as it is at the federal level or in the private sector. However, the career progression for horticulturists and botanists may not be as uniform as it is on the federal level. This phenomenon is primarily due to regional and local differences. Therefore, it is important to educate yourself about the career paths in the locality in which you are interested.

There are two resources that can help you determine the career path of botanists and horticulturists at the state and local level. Those resources are:

> *State Administrative Official: Classified by Functions*
> Council of State Governments
> Iron Works Pike
> P.O. Box 11910
> Lexington, KY 40578

This resource is organized by state and updated biennially. It also provides the names and addresses of officials in each type of state administrative department with whom you can correspond to learn about career paths for employees in your particular field of botany or horticulture.

> *Municipal Year Book*
> International City Management Association
> 777 North Capitol Street, NE
> Suite 500
> Washington, DC 20002-4201

This publication lists the addresses of city and county administration offices and departments and is updated annually. Again, it is an excellent resource for identifying professionals who can give you good insight into the career paths available in local government agencies.

MATCHING YOUR INTERESTS TO OPPORTUNITIES IN GOVERNMENT

If you have an interest in *resource management,* the U.S. Department of the Interior provides numerous opportunities for people interested in land and water reclamation. The Bureau of Reclamation is the government's principal water development and water management agency. The Geological Survey investigates and assesses land and water, conducts research on global change, and investigates natural hazards. Likewise, the Bureau of Land Management is responsible for nearly 300 million acres of public lands. The resources that are managed include such things as rangeland vegetation, endangered plant species, and conservation of wilderness areas.

If you have an interest in *promoting technical and scientific knowledge* about plants, the Smithsonian Institution is a federal agency that provides the opportunity for botanists and horticulturists to present exhibits, conduct research, publish studies, and participate in international programs of scholarly exchange.

If you have an interest in *international work* as well as an interest in horticulture or botany, the Peace Corps provides an unprecedented opportunity for you to share your knowledge with the world. The Peace Corps recruits more than 3,000 people annually to promote world peace, friendship, and understanding in over 70 developing countries. For periods of two to three years or longer, volunteers work in a variety of entry-level positions helping governments and their people learn to grow, propagate, and harvest sustainable crops to improve the quality of life in those countries. Horticulture and botany professionals are in high demand by the Peace Corps.

If you have an interest in our country's *space program,* National Aeronautics and Space Administration (NASA) offers opportunities to address such cutting-edge issues as feeding astronauts in space. NASA has been working to identify plant materials that can be used to sustain life in space. Through five university centers around the country, NASA has funded faculty to research and develop strategies for growing, processing, and preparing food in space.

Tuskegee University's NASA Center for Food Production, Processing and Waste Management in Controlled Ecological Life Support Systems (CELSS) is one such project. The CELSS Project uses the expertise of plant physiologists, plant nutritionists, plant pathologists, horticulturists, microbiologists, plant breeders, plant geneticists, plant modelers, biochemists, food chemists, food scientists, soil scientists, and water chemists as well as numerous engineers and computer scientists. The overall goal of the CELSS Project is to provide tested information and technologies applicable to bioregenerative food production systems for life support on long-term manned space missions. Specifically, the Center is developing information, computer simulated models, methodologies, and technology for sweet potato and peanut biomass production and processing. This work also includes waste management and recycling of these crops selected by NASA for CELSS.

The Center is organized into interdisciplinary teams of life scientists and engineers that work together on specific objectives and long-term goals. Integral to the goal of the Center is the development of both basic and applied research information. In addition, the Center is dedicated to the training of young scientists and engineers, especially underrepresented minorities that will increase the professional pool in these disciplines and contribute to the advancement of space sciences and exploration.

The Center serves as a focal point for integrating research on subsurface crops within the program. The research of the Center falls under one of four subsystems:

- Biomass production, which includes environmental control and computer modeling of plant growth in CELSS conditions;
- Sweet potato and peanut germplasm development for CELSS conditions;
- Nutrition and processing, which includes nutrient analysis and preparation of a variety of foods from these crops; and
- Recycling of waste from biomass production and processing.

The Center is composed of six functional working groups that interface on a continuous basis. The targeted goals of each working group within the Center are summarized below:

- Growing Systems and Environmental Factors (GRO): to ascertain the best systems and environmental conditions for growing sweet potato and peanut hydroponically for space missions.
- Microgravity Applications and Controls Group (MAC): to adapt efficient crop producing hydroponic systems for use under microgravity conditions and to design and build automated environmental control systems for crop growing and waste management systems.
- Plant Modeling Group (PAM): to establish a database from existing information, recommend and design experiments for retrieving needed information and, based on these data, produce models that effectively predict growth and yield of sweet potatoes and peanuts in CELSS.

- Nutrition and Food Processing Group (NAF): to analyze the nutrient composition of the edible parts (greens and roots/nuts) of plants grown under controlled environmental conditions and to process these parts into a variety of nutritious and palatable foods.
- Germplasm Development Group (GED): to examine the germplasm available for peanut and sweet potato; to select and breed the best lines for growing in CELSS and to further improve them through biotechnology.
- Waste Management and Recycling Group (WAM): to ascertain quantities of inedible plant biomass available from sweet potato and peanut in CELSS, to analyze the biomass for chemical composition and to establish how all organic and inorganic waste resources from hydroponic culture of these crops will be recovered and recycled in CELSS.

You can obtain additional information on the CELSS's homepage: http://www.tusk.edu/tusk/agriculture/web/pages/celss.htm

If you have an interest in *intelligence work,* numerous government agencies, such as the National Security Agency or the Central Intelligence Agency, offer an opportunity for people with backgrounds in botany and horticulture. In the intelligence world, professionals in these areas tend to work with state-of-the-art technology in the identification of indigenous plant materials around the world. This provides vital information about environmental conditions that may impact the quality of life of the world community.

SPECIAL TIPS

Participation in a college or university cooperative education (co-op) program is one of the best ways to secure a permanent position with a government agency. As a co-op student with a federal agency, you are considered to be an employee of the agency and upon graduation you are eligible to convert your co-op employment to permanent employment without competition, if full-time positions are available. The real benefit of this conversion is that your seniority is dated from your original employment date as a co-op student. This has a positive impact on your eligibility for certain federal benefits, including sick leave, vacation, and ultimately retirement. See your co-op office about applying to federal, state, and local agencies in which you are interested.

Likewise, special internship programs allow students to gain experience with a government agency prior to graduation. Internships are typically shorter in duration than co-op experiences; however, they provide good opportunities to learn about the nature of work and the career opportunities available in federal, state, and local government agencies.

In both co-op and internship programs, participants generally are selected on the basis of academic performance and demonstrated leadership skills. It is important to use the resources of your college or university in preparing to compete for these positions. A strong resume and good interviewing skills are essential in being able to secure an internship or co-op position.

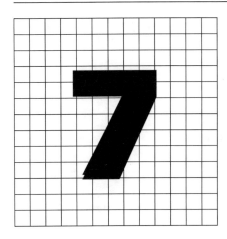

INDUSTRY OPPORTUNITIES FOR BOTANISTS AND HORTICULTURISTS

Industry provides many botanists and horticulturists with a wide and diverse range of career paths. While the opportunities are always changing in terms of how botanists and horticulturists work in industry, there is not indication that industry will lose its position as an important place of employment for some people in these career fields.

Private industry employs ten percent of botanists. The types of industries that are most likely to hire people in this field are drug companies, the oil industry, the chemical industry, lumber and paper companies, seed and nursery companies, fruit growers, food companies, fermentation industries (including breweries), biological supply houses, and biotechnology firms.

It is not necessary for botanists and horticulturists to begin their careers in industry in order to have a successful career path in the private sector. For example, some botanists and horticulturists who initially join federal or state regulatory agencies and gain several years of experience find it easy to move into industrial positions.

The following are some resources that provide employment information and opportunities for people in botany and horticulture.

SOURCES OF INFORMATION ON BOTANY AND HORTICULTURE CAREERS

Every year, many people seek to begin a career in botany and horticulture. However, many become frustrated in their job search because typical sources of job leads do not tend to list the organizations that seek people in botany and horticulture. This does not mean that such sources do not exist. There are a number of good resources that should be consulted when undertaking a search for summer jobs, internships, cooperative education opportunities, and permanent positions. These resources include, but are not limited to, the following resources:

Directory of American Research and Technology

Research Centers Directory

Directory of American Agriculture

Agricultural Research Institute Directory

Life Sciences Agencies and Organizations Directory

Directory of Extension (This directory gives the addresses of all state extension offices.)

Encyclopedia of Associations (This lists almost every professional association in existence today.)

Job Choice: Science and Engineering

Federal job listings

College and university alumni associations where biology, botany, and horticulture degrees are granted

SOME POTENTIAL EMPLOYERS OF BOTANY AND HORTICULTURE PROFESSIONALS

Following are lists of potential employers of people in botany and horticulture in agricultural, paper and lumber, and pharmaceutical industries.

Agricultural Industries

Archer Daniels Midland Co.
4666 Faries Parkway
Box 1470
Decatur, IL 62525

Birdsong Corp.
612 Madison Avenue
Suffolk, VA 23434

Castle and Cooke, Inc.
10900 Wilshire Boulevard
Los Angeles, CA 90024-6501

Conagra Inc.
One Conagra Drive
Omaha, NE 68102-5001

CPC International Inc.
International Plaza
P.O. Box 8000
Englewood Cliffs, NJ 07632 9976

Curtice Burns Foods Inc.
90 Linden Place
P.O. Box 681
Rochester, NY 14603-0682

Del Monte Foods Co.
One Market
San Francisco, CA 94105-1313

Dole Food Co. Inc.
31355 Oak Crest Drive
Westlake Village, CA 91361-4633

Duda and Sons, Inc.
1975 West State Road, No. 426
Oviedo, FL 32766

Flowers Industries Inc.
11796 U.S. Highway 19 South
P.O. Box 1338
Thomasville, GA 31792

General Mills Inc.
Number One General Mills Boulevard
Minneapolis, MN 55426

Grupo Industrial Maseca Sa de CV Gimsa
Av. Paseo de la Reforma 300, Piso 9
Mexico 6, DF
Mexico

Hamakua Sugar Co. Inc.
P.O. Box 250
Paauilo, HI 96776

Hanover Foods Corp.
York Street Extended
P.O. Box 334
Hanover, PA 17331-0334

Holly Farms Corp.
1755-D Lynnfield Road
No. 149
Memphis, TN 38187

J. M. Smucker Co.
One Strawberry Lane
Orville, OH 44667-0280

Kellogg Co.
One Kellogg Square
P.O. Box 3599
Battle Creek, MI 49016-3599

Lancaster Colony Corp.
37 West Broad Street
Columbus, OH 43215-4177

Maui Land and Pineapple Co., Inc.
Kane Street
P.O. Box 187
Kahului, HI 96732

Ernest J. Miller Enterprises
Second West and Fourth North
Hyrum, UT 84319

Montedison Spa
31 Foro Buonaparte
Milan 20121
Italy

Nabisco Inc.
7 Campus Drive
Parsippany, NJ 07054

Ore-Ida Foods, Inc.
220 West Parkcenter Boulevard
Boise, ID 83707

Pacific Tomato Growers, Inc.
503 Tenth Street
Palmetto, FL 33561

Pilgrims Pride Corp.
110 South Texas Street
P.O. Box 93
Pittsburg, TX 75686

PM Holdings Corp.
1401 South Hanley Road
St. Louis, MO 63144

Ralcorp Holdings Inc.
Suite 2900
800 Market Street
St. Louis, MO 63101

Ralston Purina Co.
Checkerboard Square
St. Louis, MO 63164-0001

Riviana Foods Inc.
2777 Allen Parkway
Houston, TX 77019

RJR Nabisco Holdings Corp.
1301 Avenue of the Americas
New York, NY 10019-6013

RJR Nabisco Inc.
1301 Avenue of the Americas
New York, NY 10019-6013

Savannah Foods and Industries Inc.
P.O. Box 339
Savannah, GA 31402-0339

Seneca Foods Corp.
1162 Pittsford-Victor Road
Pittsford, NY 14534-3818

J. R. Simplot Foods, Inc.
P.O. Box 1059
Caldwell, ID 83606

Stokely USA Inc.
1055 Corporate Center Drive
Oconomowoc, WI 53066-4829

Sunsweet Growers, Inc.
P.O. Box 232
Yuba City, CA 95991-9351

Tomkins PLC
East Putney House
84 Upper Richmond Road
London SW15 2ST
England

United Brands
250 East Fifth Street
Cincinnati, OH 45202

United Foods Inc.
Ten Pictsweet Drive
Bells, TN 38006

Universal Foods Corp.
433 East Michigan Street
Milwaukee, WI 53202-5106

**Paper and Lumber
Industries**

Abitibi Price Inc.
Suite 680
207 Queen S Quay West
Toronto, Ontario M5J 2P5
Canada

Albany International Corp.
1373 Broadway
Albany, NY 12204-2697

Alco Standard Corp.
Box 834
Valley Forge, PA 19482-0834

American Cyanamid Co.
One Cyanamid Plaza
Wayne, NJ 07470-2076

American Greetings Corp.
One American Road
Cleveland, OH 44144

Anheuser Busch Cos. Inc.
One Busch Place
St. Louis, MO 63118-1849

Asia Pacific Resources Intl. Holdings Ltd.
80 Robinson Road
Suite 15-00
Singapore 0106 Singapore

Asia Pulp and Paper Co. Ltd.
3 Shenton Way
25-01 Shenton House
Singapore 0106 Singapore

Avenor Inc.
1250 Rene Levesque Boulevard West
Montreal, Quebec H3B 4Y3
Canada

Boise Cascade Corp.
1111 West Jefferson Street
Boise, ID 83702-0000

Bowater Inc.
55 East Camperdown Way
P.O. Box 1028
Greenville, SC 29602

Champion International Corp.
One Champion Plaza
Stamford, CT 06921

Chemed Corp.
2600 Chemed Center
255 East Fifth Street
Cincinnati, OH 45202-4726

Cytec Industries Inc.
Five Garret Mountain Plaza
West Paterson, NJ 07414

Durango Industrial Group
150 Potasio
Ciudad Industrial
011 52 181 40078
Durango, Mexico

E. I. Du Pont de Nemours and Co.
1007 Market Street
Wilmington, DE 19898

Engelhard Corp.
101 Wood Avenue
Iselin, NJ 08830-2703

Federal Paper Board Co. Inc.
75 Chestnut Ridge Road
Montvale, NJ 07645-1859

Fletcher Challenge Ltd.
Fletcher Challenge House
810 Great South Road
Penrose, Auckland
New Zealand

Fort Howard Corp.
1919 South Broadway
Green Bay, WI 54304

Gibson Greetings, Inc.
2100 Section Road
Cincinnati, OH 45237-3510

Harnischfeger Industries Inc.
13400 Bishops Lane
Brookfield, WI 53005-6203

Hercules Inc.
Hercules Plaza
1313 North Market Street
Wilmington, DE 19894-0001

International Paper Co.
Two Manhattanville Road
Purchase, NY 10577-2196

James River Corp. of Virginia
120 Tredegar Street
Richmond, VA 23219-4306

J. B. Hunt Transport Services Inc.
615 J. B. Hunt Corporate Drive
Lowell, AR 72745

Jefferson Smurfit Corp.
8182 Maryland Avenue
St. Louis, MO 63105

Manville Corp.
717 17th Street
Denver, CO 80202-3330

Mead Corp.
Mead World Headquarters
Courthouse Plaza Northeast
Dayton, OH 45463-0001

Moore Corp. Ltd.
1 First Canadian Place
Toronto, Ontario M5X 1G5
Canada

Nash Finch Co.
7600 France Avenue South
P.O. Box 355
Minneapolis, MN 55435

New York Times Co.
229 West 43rd Street
New York, NY 10036-3913

News Corp. Ltd.
2 Holt Street
Sydney, New South Wales
Australia

Noranda Inc.
BCE Place, Suite 4100
181 Bay Street
P.O. Box 755
Toronto, Ontario M5J 2T3
Canada

Rykoff Sexton Inc.
1050 Warrenville Road
Lisle, IL 60532-5201

Scott Paper Co.
Scott Center
Suite 300
2650 North Military Trail
Boca Raton, FL 33431-0000

Sequa Corp.
200 Park Avenue
New York, NY 10166-0005

Sonoco Products Co.
North Second Street
P.O. Box 160
Hartsville, SC 29551-0160

St. Joe Paper Co.
Suite 400
1650 Prudential Drive
Jacksonville, FL 32207-8176

Supervalu Inc.
11840 Valley View Road
Eden Prairie, MN 55344-3643

Sweetheart Holdings Inc.
7575 South Kostner Avenue
Chicago, IL 60652

Swift Transportation Co. Inc.
1455 Hulda Way
Sparks, NV 89431-6026

Sysco Corp.
1390 Enclave Parkway
Houston, TX 77077-2099

Union Camp Corp.
1600 Valley Road
Wayne, NJ 07470-2066

Valhi Inc.
Suite 1700
5430 LBJ Freeway
Dallas, TX 75240-2697

Werner Enterprises Inc.
Interstate 80 and Highway 50
P.O. Box 37308
Omaha, NE 68137-0308

Westvaco Corp.
Westvaco Building
299 Park Avenue
New York, NY 10171-0002

Willamette Industries Inc.
Suite 3800
1300 Southwest Fifth Avenue
Portland, OR 97201

Xerox Corp.
P.O. Box 1600
Stamford, CT 06904-1600

Pharmaceutical Industries

Abbott Laboratories
100 Abbott Park Road
Abbott Park, IL 60064-3500

American Cyanamid Co.
One Cyanamid Plaza
Wayne, NJ 07470-2076

American Home Products Corp.
Five Giralda Farms
Madison, NJ 07940-0874

Baxter International Inc.
One Baxter Parkway
Deerfield, IL 60015-4633

Bristol Myers Squibb Co.
345 Park Avenue
New York, NY 10154-0004

China Industrial Group Inc.
Suite 1801
18/F Central Plaza
18 Harbour Road
Wanchai, Hong Kong

Eli Lilly and Co.
Lilly Corporate Center
Indianapolis, IN 46285-0001

Glaxo Holdings PLC
Lansdowne House
Berkeley Square
London W1X 6BQ
England

ICN Pharmaceuticals Inc.
3300 Hyland Avenue
Costa Mesa, CA 92626

Ivax Corp.
8800 Northwest 36th Street
Miami, FL 33178-2433

Marion Merrell Dow Inc.
9300 Ward Parkway
Kansas City, MO 64114-3321

Merck and Co. Inc.
One Merck Drive
P.O. Box 100
Whitehouse Station, NJ 08889-0100

Novo Nordisk AS
Novo Alle
DK-2880
Bagsvaerd
Denmark

Pfizer Inc.
235 East 42nd Street
New York, NY 10017-5755

Pharmacia Corp.
15 Frosundaviks Alle
Stockholm S-171 97
Sweden

Rhone Poulenc Rorer Inc.
500 Arcola Road
Collegeville, PA 19426-0107

Sigma Aldrich Corp.
3050 Spruce Street
St. Louis, MO 63103-2530

Smithkline Beecham PLC
New Horizons Court
Great West Road
Brentford, Middlesex TW8 9EP
England

Upjohn Co.
7000 Portage Road
Kalamazoo, MI 49001-0199

Warner Lambert Co.
201 Tabor Road
Morris Plains, NJ 07950-2693

Wellcome PLC
Unicorn House
160 Euston Road
London NW1 2BP
England

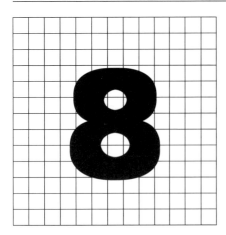

PLANNING FOR
GRADUATE SCHOOL

If your career in botany or horticulture requires an advanced degree, your selection of an appropriate graduate school is very important. First and foremost, you will want to pursue your graduate study at an institution that best prepares you for the type of career that you are seeking. However, you will also want to pursue your graduate study at an institution that best meets your personal needs in terms of such things as curriculum requirements, geographic location, and personal cost versus financial aid availability.

According to Garner (1995), top-ranked institutions usually provide the best platform for almost any career. However, admission to such programs is not enough. You must perform well and you must earn the support of your major professor in order to gain the most from your graduate education. Over the long term, the reputation of the university you select and the recommendations of your graduate faculty advisor may help open doors and may help assure your career success. Therefore, when possible and appropriate, graduate schools consistently ranked among the top 25 programs in your field are excellent schools to consider. The reputations of these institutions, the quality of their faculty, and your personal credentials will make you more competitive in a tight job market.

IDENTIFYING THE RIGHT GRADUATE SCHOOL FOR YOU

There are numerous directories available to help you identify graduate schools in your area of interest. These directories will provide information on such things as the faculty/student ratio (it should be low); the number of master's and doctoral degrees awarded annually; the percentage of women and minorities admitted; whom to contact for application materials; and the list of standardized tests that are required for admission.

There are also publications that rank graduate programs. One of these publications is *Gourman's*. It rates domestic and international graduate and professional programs. Likewise, *U.S. News and World Report* annually ranks U.S. graduate schools and specific graduate programs. These directories are useful tools when making a decision about which graduate programs you wish to apply to.

In using these rankings, it is advisable to examine the factors used to rank schools. Some factors may be more important to you than others. For example, if you are planning to obtain a Ph.D., it will be important to know how the department or school ranks among other institutions and how it ranks in terms of faculty publications. However, if you plan to enter private industry after graduation, it will be important to know how industry rates the department or school to which you plan to apply.

It is also advisable to determine if the program of study in which you are interested is ranked separately from the school or university in general. If it is, you will want to determine what the departmental ranking is in relation to the school/college or the university. A good rule of thumb when trying to make this determination is to assess whether or not the ranking of your intended department is as high or higher than the overall ranking of the school or university. For example, you would want to determine if the Horticulture Department is ranked as highly as the School/College of Agriculture and/or the University. If it is equally ranked or ranked higher, the quality of the department is probably indisputable. If it is ranked lower, you will want to ask probing questions of the faculty and graduate students at that institution before making the decision to apply and/or attend.

Another excellent resource for information about graduate programs in your area of interest is members of your undergraduate faculty. Some people who plan to go on to graduate school fail to seek the advice and guidance of undergraduate faculty when making decisions about graduate school. This can be a significant oversight. University and college faculty are generally aware of their counterparts at other universities. They are usually knowledgeable about the quality and quantity of the research of faculty at other institutions. They also tend to be aware of other faculty members' reputation in the field.

The input of faculty can be particularly helpful when you are in the process of completing the graduate school application form. Not only can these faculty provide you with letters of recommendation that address your academic performance and potential, they can also assist you in developing a well-targeted "Statement of Purpose", which is generally required on all graduate school applications.

Unlike admissions to undergraduate school, graduate applications are reviewed and evaluated by faculty members in the department in which you plan to study. Therefore, before you apply to graduate school, it is advisable to talk to the faculty in your undergraduate department and to learn all that you can about the faculty in the department to which you plan to apply. For instance, some good questions to ask might be:

Where are their degrees from?

What is their area of expertise?

What books and articles have they published?

Are their professional interests similar to yours?

VGM's *Great Jobs for Engineering Majors* (Garner, 1995) provides the following suggestions on how to assess if the graduate faculty's research interests are similar to yours. Although provided as a guide for engineering students, the advice is also appropriate for botany and horticulture students.

- Read resources such as the *Peterson's Guide.* It provides indexes of research areas and surveys of professional society research publications and journals. This resource helps to highlight the nature of research being carried on at different universities, and the faculty members doing it.

- Conduct literature searches to determine the research papers the faculty members have written and also to become more familiar with the field.

- Call the department and inquire about their active areas of research. Follow up with telephone calls to the faculty members who are active researchers and ask about their work. It is important for you to determine if their area of interest matches yours. You also want to assess whether you think you could work and study with this person. Any call like this should end with a request for information about other universities and colleagues working in this area of research.

- Visit schools and tour the department in which you are interested. This is the best means to evaluate any school or department. How do you feel when you are on the campus? In the department? Talking to the faculty? Talking to current graduate students?

Key issues in deciding where to apply for graduate study include:

- A good match between your research interests and the department's strengths;

- The reputation of the department and/or the specialization in which you are interested; and

- The overall quality of the institution.

APPLYING TO GRADUATE SCHOOL

Most graduate programs require the Graduate Record Examination (GRE). It is advisable to take the GRE during your senior year of college. Even if you do not plan to go to graduate school immediately after completing your undergraduate degree, it is good to take the GRE during your senior year. There are three principal reasons for this.

First, as an undergraduate student, you are used to taking tests and exams. Second, while you are in school, your general knowledge, as well as specific discipline knowledge, is current and up to date. Third, the scores are good for five years, and having taken the exam, all of your options for the future remain open. It can be intimidating to take this exam when you have not been in school for several years!

The next step in the process is the completion of the graduate application. The basic steps include:

- Completing an application form;
- Submitting official transcripts;
- Providing three letters of reference;
- Writing a statement of purpose; and
- Requesting financial assistance.

FINANCING YOUR GRADUATE EDUCATION

Paying for graduate school should not be a barrier to an academic career. If you have a strong record of academic achievement, you should be competitive for numerous fellowships and graduate assistantships. These not only pay your tuition and fees but also provide a monthly stipend for living expenses. Some of these sources include:

Teaching assistantships. These provide stipends, and sometimes tuition waivers, to full-time graduate students for assisting a faculty member in teaching undergraduate classes.

Research assistantships. These provide stipends, and sometimes tuition waivers, to full-time graduate students for assistance on a faculty research project.

Fellowships. These provide money to full-time graduate students to cover the costs of study and living expenses and are not based on an obligation to assist in teaching or conducting research. Some fellowships are funded by the school but others are available from outside sources, such as the National Science Foundation. Those fellowships provide an excellent credential when applying for post-doctoral positions and ultimately faculty positions because they are so competitive. If these sources of funding do not seem available to you, student loans are available to graduate students.

In seeking funding for your graduate education, it is highly recommended that you do your homework. Talk to faculty and graduate schools, and learn as much as you can about the process of competing for fellowships and assistantships. If you pre-judge your own competitiveness, you may be making a costly mistake and incurring unnecessary debt!

It is important to have a good undergraduate record and evidence of a commitment to the field. This commitment can be in the form of paid, related work experience or volunteer experience. The faculty committee that reviews applications and makes recommendations for the awarding of financial assistance needs to see that you can make a substantive contribution to their teaching and/or research.

UNIVERSITIES AND COLLEGES IN THE UNITED STATES OFFERING DEGREES IN AGRICULTURE

Alabama

Alabama A&M University
Normal, AL 35762

Auburn University
Auburn University, AL 36849

Tuskegee Institute
Tuskegee Institute, AL 36088

Alaska

University of Alaska
Agricultural Experiment Station
Palmer, AK 99645

Arizona

Arizona State University
Tempe, AZ 85281

University of Arizona
Tucson, AZ 85721

Arkansas

Southern Arkansas University
Magnolia, AR 71753

University of Arkansas
Fayetteville, AR 72701

University of Arkansas
Monticello, AR 71655

University of Arkansas
Pine Bluff, AR 71601

California

California Polytechnic State University
San Luis Obispo, CA 93407

California State Polytechnic University
Pomona, CA 91768

California State University
Chico, CA 95926

California State University
Fresno, CA 93740

Humboldt State University
Arcata, CA 95521

University of California
Davis, CA 95616

University of California
Riverside, CA 92521

Colorado

Colorado State University
Fort Collins, CO 80523

Fort Lewis College
Durango, CO 81301

Connecticut

University of Connecticut
Storrs, CT 06268

Delaware

Delaware State College
Dover, DE 19901

University of Delaware
Newark, DE 19711

Florida

Florida Southern College
Lakeland, FL 33802

University of Florida
Gainesville, FL 32611

Georgia

Abraham Baldwin Agricultural College
Tifton, GA 31794

Berry College
Mount Berry, GA 30149

Fort Valley State College
Fort Valley, GA 31030

University of Georgia
Athens, GA 30602

Hawaii

University of Hawaii
Honolulu, HI 96822

Idaho

College of South Idaho
Twin Falls, ID 83301

Ricks College
Rexburg, ID 83440

University of Idaho
Moscow, ID 83843

Illinois
Illinois State University
Normal, IL 61761

Southern Illinois University
Carbondale, IL 62901

University of Illinois
Urbana, IL 61801

Western Illinois University
Macomb, IL 61455

Indiana
Purdue University
West Lafayette, IN 47907

Iowa
Iowa State University
Ames, IA 50011

Kansas
Fort Hays State University
Hays, KS 67601

Kansas State University
Manhattan, KS 66506

McPherson College
McPherson, KS 67460

Kentucky
Morehead State University
Morehead, KY 40351

Murray State University
Murray, KY 42071

University of Kentucky
Lexington, KY 40506

Western Kentucky University
Bowling Green, KY 42101

Louisiana
Louisiana State University
Baton Rouge, LA 70803

Louisiana Tech University
Ruston, LA 71272

McNeese State University
Lake Charles, LA 70609

Nichols State University
Thibodaux, LA 70310

Northeast Louisiana University
Monroe, LA 70209

Northwestern State University
Natchitoches, LA 71497

Southeastern Louisiana University
Hammond, LA 70402

Southern University
Baton Rouge, LA 70813

University of Southwestern Louisiana
Lafayette, LA 70504

Maine
University of Maine
Orono, ME 04469

Maryland
University of Maryland
College Park, MD 20742

University of Maryland
Princess Anne, MD 21853

Massachusetts
University of Massachusetts
Amherst, MA 01003

Michigan
Michigan State University
East Lansing, MI 48824

Michigan Technological University
Hancock, MI 49930

Northern Michigan University
Marquette, MI 49855

Minnesota
University of Minnesota
St. Paul, MN 55108

University of Minnesota Technical College
Crookston, MN 56716

University of Minnesota Technical College
Waseca, MN 56093

Mississippi
Alcorn State University
Lorman, MS 39096

Mississippi State University
Mississippi State, MS 39762

Missouri
Central Missouri State University
Warrensburg, MO 64093

Lincoln University
Jefferson City, MO 65101

Missouri Western State College
St. Joseph, MO 64507

Northeast Missouri State University
Kirksville, MO 63501

Northwest Missouri State University
Maryville, MO 64468

Southwest Missouri State University
Springfield, MO 65802

University of Missouri
Columbia, MO 65211

Montana
Montana State University
Bozeman, MT 59717

Northern Montana College
Havre, MT 59501

Nebraska
University of Nebraska
Lincoln, NE 68583

Nevada
University of Nevada
Reno, NV 89557

New Hampshire
University of New Hampshire
Durham, NH 03824

New Jersey
Rutgers University
New Brunswick, NJ 08903

New Mexico
New Mexico State University
Las Cruces, NM 88003

New York
Cornell University
Ithaca, NY 14853

North Carolina
North Carolina A&T State University
Greensboro, NC 27411

North Carolina State University
Raleigh, NC 27650

North Dakota
North Dakota State University
Fargo, ND 58105

Ohio
Ohio State University
Columbus, OH 43210

Wilmington College
Wilmington, OH 45177

Oklahoma
Cameron University
Lawton, OK 73505

Langston University
Langston, OK 73050

Oklahoma Panhandle State University
Goodwell, OK 73939

Oklahoma State University
Stillwater, OK 74074

Oregon
Oregon State University
Corvallis, OR 97331

Pennsylvania
Delaware Valley College of Science and Agriculture
Doylestown, PA 18901

Pennsylvania State University
University Park, PA 16802

Temple University
Ambler, PA 19002

Puerto Rico
University of Puerto Rico
Mayaguez, PR 00708

University of Puerto Rico
Rio Piedras, PR 00928

Rhode Island
University of Rhode Island
Kingston, RI 02881

South Carolina
Clemson University
Clemson, SC 29631

South Dakota
South Dakota State University
Brookings, SD 57007

Tennessee
Austin Peay State University
Clarksville, TN 37040

Middle Tennessee State University
Murfreesboro, TN 37130

Tennessee Technological University
Cookeville, TN 38505

University of Tennessee
Knoxville, TN 37901

University of Tennessee
Martin, TN 38238

Texas
Abilene Christian University
Abilene, TX 79601

East Texas State University
Commerce, TX 75428

Prairie View A&M University
Prairie View, TX 77445

Sam Houston State University
Huntsville, TX 77341

Southwest Texas State University
San Marcos, TX 78666

Stephen F. Austin State University
Nacogdoches, TX 75962

Texas A&I University
Kingsville, TX 78363

Texas A&M University
College Station, TX 77843

Texas Tech University
Lubbock, TX 79409

West Texas State University
Canyon, TX 79015

Utah

Brigham Young University
Provo, UT 84602

Utah State University
Logan, UT 84322

Vermont

University of Vermont
Burlington, VT 05405

Virginia

Old Dominion University
Norfolk, VA 23508

Virginia Polytechnic Institute and State University
Blacksburg, VA 24061

Virginia State University
Petersburg, VA 23806

Washington

University of Washington
Seattle, WA 98195

Washington State University
Pullman, WA 99164

West Virginia

West Virginia University
Morgantown, WV 26506

Wisconsin

University of Wisconsin
Green Bay, WI 54302

MENTORING TIPS FROM PEOPLE IN THE FIELDS OF BOTANY AND HORTICULTURE

In many areas of botany and horticulture it is difficult to readily identify people who are working in the various occupational fields associated with these two career areas. Botany and horticulture are not like accounting, law, engineering, or teaching, where numerous people in the community have prepared for and are pursuing these occupations. Where would one go to ask a question or obtain advice?

This section has been designed to provide an opportunity to hear directly from professionals in the various branches and occupational fields of botany or horticulture. Professionals from different specialties and locations were surveyed, and each was asked to describe three factors that would be helpful to the readers of this book. Specifically, the respondents were asked to explain

- Why they chose their career field
- What tasks they perform on a regular basis
- What tips they would give to people considering their occupational field

The following are their thoughts and advice.

FROM LISA, A TECHNICIAN IN PLANT SCIENCE

Why I Chose Plant Science

With definite scientific interests, I chose plant science because many other branches of research require the use/killing of animals. I find research preferable to industry because it offers the opportunity to continuously learn new things. In terms of career advancement, research has more teamwork and less competition (theoretically). Applied research has the added benefit that the value of your work is more immediately tangible.

What I Do

I do research on various crops to try to solve growers' immediate problems and to try to provide an explanation for how systems work. Answers I get may be immediately useful to a grower or they may provide a basis for further research. I have been involved with genetic crosses, variety trials, and detailed examination of a biocontrol fungus. My daily duties can range from cleaning a greenhouse to harvesting a field to chemiluminescence. One of the best parts of the job is the great variety of things I get to do. My work involves greenhouse work, laboratory testing, fieldwork, and computer analysis of data. As new experiments arise, I am exposed to many different types of equipment such as microscopes, centrifuges, autoclaves, electrophoretic equipment, spectrophotometers, planters, rototillers, growth chambers, and sometimes unique equipment designed just for the experiment we are working on.

Tips for People Considering This Field

After taking the basic classes required for a major in biology or botany, there will be room for several electives. Take classes that you're interested in rather than concentrating on marketable classes. While it is good to have classes that will sell you to your first employer, you never know what field will be most employable in a few years. It is best to try to make the extra effort to find a job that will be fun for you.

Summer jobs doing greenhouse work would be helpful. An independent study project is a good way to get research experience. A less formal approach is to ask a favorite professor if you can volunteer to help them on a project. What might be "grunt work" for them could be a valuable experience for you.

FROM NORMAN, A PLANT GENETICIST AND ASSOCIATE PROFESSOR

Actually, I am not a very good person to ask for such advice. I sort of "backed" into my career. I never really thought that I wanted to be a "horticulturist." I just followed my interests and the job market and ended up as a plant geneticist working with horticultural crops.

I received a bachelor of science degree in chemistry and about the same time decided I was more interested in biological systems. As a result, I obtained a master's degree in biology (studying stress in rabbit populations).

My first job was with an engineering firm. I was hired to write environmental impact statements. Later, I worked for another civil engineering firm as supervisor of their water quality lab. In this job I found that I was back to chemistry! Somewhere in that period I also wrote a book on the flora of the Sierra Nevada.

Finally, I decided to go back to school and get a Ph.D. in botany. It turned out that the genetics program was more to my liking. I graduated in plant genetics with an emphasis in evolution. After receiving my Ph.D., I secured a post-doctoral position in plant biochemistry/evolution. When I finished that post-doc experience, I accepted my current faculty position (which was originally defined as a plant biochemist working with seeds but which since has been redefined—due to my interests and developments in the field of plant science—as a plant geneticist working with breeders).

My advice would be to follow one's interests, with at least a little attention paid to possible alternative employment opportunities.

FROM HELENE, A PLANT PATHOLOGIST, ASSOCIATE PROFESSOR, AND EXTENSION FACULTY MEMBER

I am an associate professor of plant pathology and my specialty is in vegetable pathology. I have a split appointment—50 percent research and 50 percent extension and outreach on diseases of vegetables. I work primarily with the commercial agriculture community.

Why I Chose This Profession

I majored in biology of natural resources as an undergraduate at the University of California at Berkeley. I always enjoyed the outdoors and was interested in a career in forestry or with the National Park Service. I went on for a master's degree in Soil Science and concentrated on plant nutrition. I took the civil service exams to see if I could make it into the forestry scene. I quickly found out that without "veterans points" (the points given to people who have served in the U.S. military) I couldn't score high enough, even though my scores, without the extra points, were in the 90s!

I started looking at careers in agriculture. Plant pathology had been my absolute favorite course as an undergraduate student. So I went with my heart. I entered a Ph.D. program in plant pathology and had the opportunity to work for a famous vegetable pathologist. My training in his lab was priceless. When I graduated, I immediately landed a faculty position at Cornell University.

What I Do

I work with farmers to help them grow healthy vegetables that can be canned, frozen, or put out as fresh produce in grocery stores. I identify the bacteria and fungi that cause disease. I look at ways to stop diseases that have started but I have a heavy focus on stopping diseases before they start. I work with fungi primarily. I study their interactions with plants. In order to do this I need special equipment such as scanning electron microscopes and pure air stations created by laminar flow hoods.

Walking into a grocery store after working with the produce industry is an enlightening and rewarding experience. You learn very quickly how difficult it is to have all the produce and extensive selection we have in the United States.

Recommendations for Others

Read as much as you can across all disciplines. If you think you are interested in a subject area, apply to work in the laboratory. I started out washing dishes in a soil scientist's laboratory and learned quite a bit about the way a scientific study is conducted. My next work study job was to grow strawberries in a greenhouse for a person studying plant nutrition. I realized I was "doing chemistry" without taking a boring course—and that made me work harder in the chemistry courses that I did take.

As a university professor, I hire high school and college students on a part-time basis just to give them exposure to what is going on. During the summer months, they are hired on full time. They learn a lot and many have gone on to good paying jobs in the sciences.

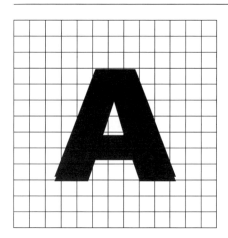

PROFESSIONAL ASSOCIATIONS IN BOTANY AND HORTICULTURE

PROFESSIONAL ASSOCIATIONS

American Association of Botanical
Gardens and Arboreta (AABGA)
786 Church Road
Wayne, PA 19087

American Association of
Nurserymen
1250 I Street, N.W., Suite 500
Washington, DC 20005

American Farm Bureau Federation
225 Touhy Avenue
Park Ridge, IL 60068

American Farmland Trust
1920 N Street, N.W., Suite 400
Washington, DC 20036

American Floral Art School
529 South Wabash Avenue, #600
Chicago, IL 60605-1679

American Floral Services
P.O. Box 12309
Oklahoma City, OK 73157

American Florists Association
2525 Heathcliff
Reston, VA 22091

American Horticultural Therapy
Association
362A Christopher Avenue
Gaithersburg, MD 20879

American Institute of Biological
Sciences
730 11th Street, N.W.
Washington, DC 20001-4584

American Institute of Floral
Designers
720 Light Street
Baltimore, MD 21230-3816

Associated Landscape Contractors
of America, Inc.
405 N. Washington Street, #104
Falls Church, VA 22046

American Landscape Horticulture
Association
2509 E. Thousand Oaks Boulevard,
Suite #109
Westlake Village, CA 91362

American Phytopathological Society
3340 Pilot Knob Road
St. Paul, Minnesota 55121
"Careers in Plant Pathology"

American Society of Agricultural
    Engineers
2950 Niles Road
St. Joseph, MI 49085-9659

American Society of Consulting
    Arborists
5130 W. 101st Circle
Westminster, CO 80030

American Society of Farm Managers
    and Rural Appraisers
950 South Cherry Street, Suite 106
Denver, CO 80222

American Society for Horticultural
    Science
701 N. Saint Asaph Street
Alexandria, VA 22314-1998
"Careers in Horticulture"

American Society of Landscape
    Architects
4401 Connecticut Avenue N.W.,
    5th Floor
Washington, DC 20008-2302

American Society of Plant
    Physiologists
15501-A Monona Drive
Rockville, MD 20955-2768
"Careers in Plant Physiology"

American Society of Plant
    Taxonomists
Rancho Santa Ana Botanic Gardens
1500 North College Avenue
Claremont, CA 91711

Association of Professional
    Landscape Designers
P.O. Box 134
Kensington, MD 20895

Botanical Society of America
Department of Botany
Ohio State University
1735 Neil Avenue
Columbus, OH 43210

Cooperative Extension Service
National Office
USDA; ES
14th Street and Independence Avenue
Washington, DC 20250

Council of Landscape Architectural
    Registration Boards
12700 Fair Lakes Circle, Suite 110
Fairfax, VA 22033

Council of Tree and Landscape
    Appraisers
1250 I Street, N.W., Suite 504
Washington, DC 20005

Crop Society of America
677 South Segoe Road
Madison, WI 53711

Ecological Society of America
2010 Massachusetts Avenue, N.W.
Suite 400
Washington, DC 20036
"Careers in Ecology"

Friends of Horticultural Therapy
362A Christopher Avenue
Gaithersburg, MD 20879

Institute for Alternative Agriculture
9200 Edmonton Road, Suite 117
Greenbelt, MD 20770

Institute for Food Technologists
221 N. LaSalle Street, Suite 300
Chicago, IL 60601

International Society of Arboriculture
P.O. Box GG
Savoy, IL 61874

Mycological Society of America
and
Phycological Society of America
P.O. Box 1897
Lawrence, KS 66044-8897

National Arbor Day Foundation/
  Institute
100 Arbor Avenue
Nebraska City, NE 68410

National Arborist Association
P.O. Box 1094
Amherst, NH 03031-1094

National Association of State
  Departments of Agriculture
1616 H Street, N.W.
Washington, DC 20006

National Farmers Union
10065 E. Harvard Avenue
Denver, CO 80231

National Future Farmers of America
  Organization
P.O. Box 15160
National FFA Center
Alexandria, VA 22309

National Landscape Association
1250 I Street, N.W. #500
Washington, DC 20005

National Wildflower Research Center
2600 FM 973 North
Austin, TX 78725-4201

North American Farm Alliance
P.O. Box 2502
Ames, IA 50010

Northeast Organic Farming
  Association (NOFA)
P.O. Box 21
South Butler, NY 13154

Office of Opportunities in Sciences
1776 Massachusetts Avenue, N.W.
Washington, DC 20036

Organization of Biological Field
  Stations
Tyson Research Center
P.O. Box 258
Eureka, MO 63025

Professional Grounds Management
  Society
10402 Ridgland Road, Suite 4
Hunt Valley, MD 21030

Sea Education Association
P.O. Box 6
Woods Hole, MA 02543

Society of American Florists
1601 Duke Street
Alexandria, VA 22314-3406

STATE NURSERY ASSOCIATIONS

Alabama
  Alabama Nurserymen's
    Association
  P.O. Box 9
  Auburn, AL 36830

Arizona
  Arizona Nurserymen's
    Association
  444 West Camelback Road,
    Suite 302
  Phoenix, AZ 85013

Arkansas
  Arkansas Nurserymen's
    Association
  P.O. Box 55295
  Little Rock, AR 72225

California
  California Association of
    Nurserymen
  1419 21st Street
  Sacramento, CA 95814

Colorado

Colorado Nurserymen's
  Association
746 Riverside Drive, Box 2676
Lyons, CO 80540

Connecticut

Connecticut Nurserymen's
  Association
24 West Road, Suite 53
Vernon, CT 06066

Delaware

Delaware Association of
  Nurserymen
Plant Science Dept.
University of Delaware
Newark, DE 19717

Florida

Florida Foliage Association
P.O. Box Y
Apopka, FL 32703

Florida Nurserymen's and
  Growers Association
5401 Kirkman Road #650
Orlando, FL 32819

Georgia

Georgia Nurserymen's
  Association
190 Springtree Road
Athens, GA 30605

Hawaii

Hawaii Association of
  Nurserymen
P.O. Box 293
Honolulu, HI 96809

Idaho

Idaho Nursery Association
1615 North Woodruff
Idaho Falls, ID 83401

Illinois

Illinois State Nurserymen's
  Association
Springfield Hilton #1702
Springfield, IL 62701

Indiana

Indiana Association of
  Nurserymen
202E 650N
West Lafayette, IN 47906

Iowa

Iowa Nurserymen's Association
7261 NW 21st Street
Ankeny, IA 50021

Kansas

Kansas Nurserymen's Association
Blueville Nursery, Route 1
Manhattan, KS 66502

Kentucky

Kentucky Nurserymen's
  Association
701 Baxter Avenue
Louisville, KY 40204

Louisiana

Louisiana Association of
  Nurserymen
4560 Essen Lane
Baton Rouge, LA 70809

Maine

Maine Nurserymen's Association
Plant and Soil Department
SMVTI, Fort Road
South Portland, ME 04106

Maryland

Maryland Nurserymen's
  Association
2 Troon Court
Baltimore, MD 21236

Massachusetts
Massachusetts Nurserymen's
Association
715 Boylston Street
Boston, MA 02116

Michigan
Michigan Association of
Nurserymen
819 North Washington Avenue,
Suite 2
Lansing, MI 48906

Minnesota
Minnesota Nursery and
Landscape Association
P.O. Box 13307
St. Paul, MN 55113

Mississippi
Mississippi Nurserymen's
Association
P.O. Box 5385
Mississippi State, MS 39762

Missouri
Missouri Association of
Nurserymen
7911 Spring Valley Road
Raytown, MO 64138

Montana
Montana Association of
Nurserymen
P.O. Box 1871
Bozeman, MT 59715

Nebraska
Nebraska Nurserymen's
Association
P.O. Box 80117
Lincoln, NE 68501

Nevada
Nevada Nurserymen's
Association
651 Avenue B
Boulder City, NV 89005

New Hampshire
New Hampshire Plant Growers
Association
194 Rumford Street
Concord, NH 03440

New Jersey
New Jersey Association of
Nurserymen
65 South Main Street
Building H, Suite 2
Pennington, NJ 08534

New Mexico
New Mexico Association of
Nurserymen
P.O. Box 667
Eslancia, NM 87106

New York
New York Nurserymen's
Association
P.O. Box 5185
Albany, NY 12205

North Carolina
North Carolina Nurserymen's
Association
P.O. Box 400
Knightdale, NC 27545

North Dakota
North Dakota Nursery and
Greenhouse Association
P.O. Box 2601
Bismark, ND 58502

Ohio
Ohio Nurserymen's Association
2021 E. Dublin-Granville Road
#185
Columbus, OH 43229

Oklahoma
Oklahoma Nurserymen's
Association
400 North Portland
Oklahoma City, OK 73107

Oregon
Oregon Association of
  Nurserymen
2780 S.E. Harrison Suite 204
Milwaukie, OR 97222

Pennsylvania
Pennsylvania Nurserymen's
  Association
1924 North Second Street
Harrisburg, PA 17102

Rhode Island
Rhode Island Nurserymen's
  Association
P.O. Box 515
North Scituate, RI 02857

South Carolina
South Carolina Nurserymen's
  Association
809 Sunset Drive
Greenwood, SC 29646

South Dakota
South Dakota Nurserymen's
  Association
3401 E. 10th Street
Sioux Falls, SD 57103

Tennessee
Tennessee Nurserymen's
  Association
P.O. Box 57
McMinnville, TN 37110

Texas
Texas Association of Nurserymen
7730 South I—H 35
Austin, TX 78745

Utah
Utah Association of Nurserymen
3500 South 9th East
Salt Lake City, UT 84106

Vermont
Vermont Plantsman's Association
P.O. Box 438
Windsor, VT 05089

Virginia
Virginia Nurserymen's
  Association
R.R. #4, Box 356
Christianburg, VA 24073

Washington
Washington State Nurserymen's
  Association
P.O. Box 670
Sumner, WA 98390

West Virginia
West Virginia Nurserymen's
  Association
Route 1, Box 33
Talcott, WV 24981

Wisconsin
Wisconsin Nurserymen's
  Association
Route 1, Box 377
Lake Mills, WI 53551